AF598278

SEED HEALTH TESTING

Seed Health Testing
Progress towards the 21st Century

Edited by

J.D. Hutchins and J.C. Reeves
National Institute of Agricultural Botany
Cambridge
UK

CAB INTERNATIONAL

CAB INTERNATIONAL
Wallingford
Oxon OX10 8DE
UK

Tel: +44 (0)1491 832111
Fax: +44 (0)1491 833508
E-mail: cabi@cabi.org

CAB INTERNATIONAL
198 Madison Avenue
New York, NY 10016-4314
USA

Tel: +1 212 726 6490
Fax: +1 212 686 7993
E-mail: cabi-nao@cabi.org

A catalogue record for this book is available from the British Library, London, UK

Library of Congress Cataloging-in-Publication Data

Seed health testing: progress towards the 21st century/edited by J.D. Hutchins and J.C. Reeves.

p. cm.

Based on papers developed from the second International Seed Testing Association (ISTA)-Plant Disease Committee symposium held in Cambridge, August 1996.

Includes index.

ISBN 0-85199-179-3 (alk. paper)

1. Seeds—Testing—Congresses. 2. Seed-borne plant diseases—Congresses. 3. Seed technology—Congresses. I. Hutchins, J.D. II. Reeves, J.C. III. International Seed Testing Association. Plant Disease Committee.

SB117.S37 1997

631.5′21—dc21 97-14634 CIP

ISBN 0 85199 179 3

Typeset in Photina by AMA Graphics Ltd, UK

Printed and bound in the UK by Biddles Ltd, Guildford and Kings' Lynn

Contents

SECTION III

SECTION IV

Introduction

This book records the proceedings of the Second International Seed Testing Association Plant Disease Committee (ISTA-PDC) Symposium held in Cambridge, UK, 5–8 August 1996. The ISTA-PDC has a long history of organizing meetings, workshops and seminars, which is briefly described by its current chairman Dr C.J. Langerak in this volume; its main purpose has been to focus attention on the importance of various aspects of seed health. Much effort has gone into furthering the objectives of the ISTA-PDC over past years but the need for continued attention to the problems caused by seedborne pathogens remains. Indeed as the growth of commercial seed trading continues, the potential for the spread of diseases which use seed as a vector assumes greater importance.

When we were asked to organize this symposium it seemed important to us to take stock of the current and emerging problems of seed health and to examine how these were being tackled on an international level. By doing so it was hoped that future implications would be identified as the end of this century approaches. Hence we considered our rather portentous choice of title for the symposium justified.

To achieve these objectives the symposium was organized into five sessions and these are reflected in the sections of this book. The first session drew on the experience of national delegates and members of the ISTA-PDC, to review new and priority seedborne diseases in major geographical areas. This was a scene-setting exercise and against this the second session attempted to examine national and international policy with regard to seed health and how this might be changing. This glimpse of the future was also reflected in the third session, which concentrated on technical developments in the methods by which seedborne pathogens are detected. In this field there have been many recent advances arising from the use of improved antisera and the polymerase chain reaction.

The fourth and fifth sessions echoed themes from the First ISTA-PDC Symposium (Ottawa, Canada, 9–11 August 1993) concentrating on the evaluation of comparative tests conducted between seed health testing laboratories and emphasizing developments in quality assurance in seed health testing. These are both matters of continuing central concern to ISTA.

A further and by no means secondary objective of the symposium was to provide a platform for the International Seed Health Initiative (ISHI) (see Chapter 12) and a forum for discussions between the ISTA-PDC and ISHI. This was an important initiative due to the complementary objectives of these organizations and the need for a coordinated approach to common problems. The groundwork for such cooperation was established in a satellite meeting during the symposium and benefits will accrue as the relationship between these organizations develops.

It was gratifying for the organizers that throughout the symposium common themes and problems began to emerge. It soon became clear that changes in approaches to seed treatment, stimulated by the move away from cheap organomercury fungicides for environmental reasons, are of common interest. Routine prophylactic seed treatment is also in question. These changes highlight the need for a more targeted seed treatment policy to reduce both financial and environmental costs. In turn, this highlights the need for reliable and timely data on the seed health status of a seed lot before rational management decisions can be taken. Consequently the development and use of rapid, sensitive, accurate and cheap test methods are paramount. For successful commerce, standardized and internationally accepted test methods are required to avoid repetitive, costly and unjustifiable retesting of seed lots destined for national and international trade. Results obtained from seed health tests should be both reliable and reproducible irrespective of which laboratory has conducted the test. Development of appropriate quality systems should provide such reassurances to customers of the seed industries, whether they are commercial or regulatory authorities.

It is clear that the issues surrounding the unquestionable importance of clean seed as a prerequisite for healthy crops are of continuing and developing significance. The key factors were covered during this symposium and we hope that their publication in this form brings them to the attention of a wider audience with the consideration that they continue to deserve.

Our thanks must be expressed to all those people who helped us in the organization and running of the symposium and in the production of this book. In particular our sincere thanks must be given to Drs Ruud van den Bulk and Kees Langerak (CPRO-DLO, Wageningen, The Netherlands) and to the staff at NIAB, especially Cathy Yates and those in the Molecular Biology and Diagnostics Section.

Dr John D. Hutchins
Dr James C. Reeves

National Institute of Agricultural Botany, Cambridge
10 February 1997

Presidential Address

I am delighted to be given the opportunity of making the introductory address to this symposium but at the same time I am aware that I am treading on dangerous ground in that I am not a seed pathologist. I hope you will bear with me though; it is always useful to get a view from the other side. This is also a chance for me to give you the views of the ISTA Executive Committee on some of the issues under consideration at this meeting.

The Plant Disease Committee (PDC) has always been one of the most active of ISTA's Technical Committees. The list of its activities is very long, and includes method development, the production of Working Sheets, the adoption of ISTA Rules for seed health testing, workshops, training courses, seminars and symposia. All of these activities help to support the main aims of ISTA which are to provide test methods for seed moving in international trade, to carry out appropriate seed-orientated research and development and to participate in training and education programmes.

In briefly reviewing the past activities of the PDC, I would say that the main emphasis of its work has been on training and education and on the development of methods for the detection of seedborne diseases on seed lots. Both of these activities have reflected the concerns of the time.

There was a need to make people aware of the importance of seed health alongside the traditional quality criteria of purity and germination and this was done through workshops and training courses. Extensive information about seedborne diseases was brought together in the *Annotated List of Seedborne Diseases*, an invaluable reference book for aspiring seed pathologists. Much of the impetus for this educational work was provided by Paul Neergaard, an influential former chairman of the PDC, aided and abetted by a formidable lady from Scotland called Mary Noble. It has continued to be a strength of the Committee and a good advertisement for ISTA. Work on test methods over the years has resulted in the production of a range of Working Sheets setting out practical methods for disease testing. That more or less brings us up to date but you will notice that I have not mentioned the International Rules for Seed Testing yet. I will come to them in a minute.

I am very pleased that the PDC has taken the initiative of organizing this meeting and it has the wholehearted support of my colleagues on the Executive Committee. To my mind we are at a turning point with regard to seed health and this will have a considerable impact on seed trading and seed testing in the future.

The blanket prophylactic use of seed treatments, for example to control seedborne diseases in cereals, is increasingly coming under scrutiny. This, side by side with continuing, if low level, interest in organic farming, has brought the role of routine testing for disease to greater prominence. Not only do we need to know whether there are diseases present, but also we need as accurate a measure as possible of the extent of infection and the likely consequence of sowing untreated seed. Some of these problems will be dealt with in the first session today.

Although pressures such as these are bound to have an effect, the introduction of new trading arrangements under the umbrella of the World Trade Organization (formerly GATT) could have an even bigger impact. As I understand it, it could become much more difficult to use plant quarantine procedures as a way of controlling seed imports into a country. It may no longer be acceptable to take the line that the absence of a pathogen in an importing country is sufficient reason to refuse entry of a lot. It might also be necessary to demonstrate in addition that there would be a negative economic outcome, for example the possibility of severe crop failures or the need for costly control programmes, if the pathogen became endemic. In other words, the emphasis would be on maximum tolerance levels for disease rather than outright rejection of seed lots. The consequence of this is that there might be more trading of seed that has low-level infection and seed testers would have to develop more tests that give good estimates of infection rates. I hope some of this will come out in the discussions on Tuesday morning in Session 2.

This brings me back to the International Rules for Seed Testing. I was once Chairman of the Rules Committee and during my time in office, there was a steady trickle of new seed health testing methods coming forward for inclusion in the Rules. This trickle has remained just that. Each time the Rules are amended a few more methods are added. The trickle has not yet become a flood and yet I have just highlighted a couple of areas where there may be a future need for standardized seed health tests. It is my view, and I think it is shared by a number of ISTA colleagues, that a major challenge for the PDC is to develop new seed health testing methods, suitable for inclusion in the Rules, as rapidly as possible. The addition of new disease testing methods to the Rules would make it possible to put all the test results for one seed lot on one certificate, which must be beneficial to seed trading. It would make the Orange and Green certificates even more useful as plant passports than they currently are.

I think challenge is the right word to describe what is needed. I suspect that the reason for the slow progress in this area is that it is more difficult to develop rules for seed health testing than it is for some other areas of seed testing. We want reliable and reproducible tests which cost the customer as little as possible, so we are aiming for the tens of dollars per test range rather than the hundreds of dollars. This is a tall order and I think beyond the resources of ISTA on its own, relying as it does on the voluntary efforts of its member scientists. ISTA cannot afford to pay for this work to be done and our Government sponsors, collectively round the world, are not exactly falling over themselves to support it. This is why a collaboration such as the International Seed Health Initiative, which we will hear more about this morning, is so valuable. The combined efforts of ISTA and seed industry scientists will, I hope, lead to a more rapid development of suitable test methods for those diseases which are a problem in the international seed trade. I would like to emphasize that this is not just my view. It also has the strong support of the Executive Committee. There will be a chance, though, for you to put your own views on the International Seed Health Initiative, at the Forum on Thursday. Audience participation is essential if we are to have a good discussion, so please come prepared to join in.

A large part of this symposium, in Session 3, is, quite rightly, devoted to discussions of progress on test methods. I would guess that the most exciting developments are those in molecular biology but this is only one weapon in our armoury and I am pleased to see included a number of papers on more traditional approaches. It is also good to see that comparative testing is considered of sufficient value to merit a session on its own in Session 4. Method development is our bread and butter but please bear in mind what I said about the Rules. Promising methods need to be turned into rules as quickly as possible.

Quality assurance (QA) is the flavour of this decade. Whatever we as individuals might think about it, it permeates our existence. It has political support in many countries because it seems to offer a way to facilitate the privatization of government functions without apparent loss of quality of service. For those of us in ISTA who work directly or indirectly for national governments this is the down side. However, there is also an up side and I would be keen to emphasize this. Properly applied, QA is an important management tool in a laboratory. It also provides benchmarks against which we can measure our capabilities. And, from the point of view of our customers, it provides assurance that the laboratory testing their seed is competent. The trend to QA was spotted by ISTA more than 10 years ago when Sandie Ednie was President. We have developed our QA systems gradually since then, to the point where we are about to start auditing our member stations. I hope that the session on QA will provide a focus for discussion of this topic and that you will benefit from it.

In conclusion, ladies and gentlemen, I would like to take this opportunity, on behalf of ISTA to thank the Director and Staff of the NIAB, particularly Dr James Reeves and his colleagues, for organizing this Symposium. They have done a tremendous job. I hope you enjoy your stay in Cambridge. I would urge you to try and find some time to look round this historic and beautiful city and, if you have not done so already, you must try punting. You cannot say that you have really been to Cambridge until you have fallen into the river.

This is also an opportunity for me, again on behalf of ISTA, publicly to acknowledge the contribution that the Chairman, full members and Working Group members of the PDC make in developing all the different facets of seed health testing. This is a very important Committee and I would like you all to know that your efforts are widely appreciated. I would particularly like to acknowledge, at this time, the work of Dr S.B. Mathur of the Danish Institute of Seed Pathology, who has given many years of valuable service to the Committee and who has had to resign as a member for family reasons.

Finally thank you all for coming to this meeting.

Mr Simon Cooper
President of the International Seed Testing Association

Fitzwilliam College, Cambridge
5 August 1996

New and Priority Seedborne Diseases in Western Europe

1

V. Cockerell

Official Seed Testing Station for Scotland, SASA, East Craigs, Edinburgh EH12 8NJ, UK

INTRODUCTION

To determine which seedborne diseases were considered of practical importance today, a questionnaire was sent to seed pathologists in Belgium, England and Wales, France, Germany, Ireland, The Netherlands, Northern Ireland and Scotland. A number of seed industry representatives were also contacted. Respondents were asked to name up to three seedborne pathogens of current economic importance in their country today and assess each pathogen as belonging to either Category A (a pathogen that has reccurred after being previously controlled) or Category B (the emergence of a new, potentially yield-limiting pathogen to their area). They were also asked to provide further information on each of the pathogens listed by answering the following specific questions.

1. What are the potential/estimated economic losses due to the pathogen?
2. Is seed testing vital to control the pathogen?
3. Are seed testing procedures adequate in terms of: identifying differences between infected seed lots; sensitivity of the test; test repeatability; time taken to produce a result; and the cost of the test?

RESULTS AND DISCUSSION

Fifteen pathogen–host combinations were recorded (Table 1.1). All pathogens listed were fungi and cereal crops were named as the host for 67% of those

Table 1.1. New and priority seedborne pathogens in Western Europe.[a]

Pathogen	Host	Country/countries
Category A		
Ascochyta spp.	pea	France
Fusarium sporotrichioides	barley	Germany
Microdochium nivale	wheat	UK and Ireland
	barley	Ireland
Pyrenophora graminea	barley	UK (Ireland)
Pyrenophora teres	barley	Ireland
Tilletia caries	wheat	Germany and UK
(*Fusarium* spp.)	(pea)	(UK)
(*Phoma* spp.)	(linseed)	(UK)
Category B		
Claviceps purpurea	rye and triticale	Germany
Microdochium nivale	oats	Ireland
	winter barley	Germany
Sphacelotheca reiliana	maize	Germany
Plasmopara halstedii	sunflower	France
(*Diaporthe helianthi*)	(sunflower)	(France)

[a]Additional pathogens listed by the seed industry as important but not listed by the pathologists are given in brackets.

recorded. Pathogen–host combinations were usually unique problems for individual countries with the exception of *Microdochium nivale* of wheat and barley, *Pyrenophora graminea* of barley and *Tilletia caries* of wheat.

Seedborne Pathogens

Pathogens listed under Category A were considered important because:

- The fungi had become resistant to the seed treatment fungicide most commonly used to control them (*P. graminea* and organomercury; *Ascochyta* spp. and benzimidazole fungicides).
- New fungicides are not as effective at controlling the disease (*M. nivale*, *Fusarium* spp. and *T. caries*).
- Seed is sown untreated either for economic or ecological reasons (*T. caries*).
- More favourable conditions for the fungus have existed in recent years (*Fusarium sporotrichioides*, *Phoma* spp.).
- There was a combination of fewer resistant varieties and less effective seed treatments (*Pyrenophora teres*).

In relation to the emergence of new pathogens, i.e. Category B:

- *Claviceps purpurea* and *M. nivale* in Germany were thought to be associated with climatic change, consecutive years of long cool and wet summers being favourable to both of these fungi.

- Three diseases were thought to have been introduced from infected seed stocks: two new races of *Plasmopara halstedii* in sunflowers (France); *Sphacelotheca reiliana* of maize (Germany); and *Diaporthe helianthi* of sunflowers (France).
- Infected stem bases arising from infected seed are the possible infection source for *M. nivale* var. *nivale*, a new leaf axil and sheath blotch disease of oats in Ireland (Diamond *et al.*, 1995).

In 1989 in the UK, strains of *P. graminea* (barley leaf stripe) were confirmed as being resistant to organomercury seed treatments (Jones *et al.*, 1989) and *P. graminea* was estimated to have caused £2 million in lost yield for Scottish spring barley growers in 1990 (Cockerell *et al.*, 1995). Infection decreased between 1990 and 1992 following the introduction of a voluntary Code of Practice in 1991 and a move from organomercury-based fungicides to effective alternatives, principally imazalil. By 1994 Cockerell and Rennie (1996) reported that the pathogen was not at levels that would effect yield but there was sufficient inoculum for the disease to increase if uncontrolled. Similarly in France, *P. graminea* was described as being of low economic importance whilst controlled within seed production systems (Cailliez, 1994).

Molinero *et al.* (1993) confirmed resistance to the benzimidazole group of fungicides (benomyl, carbendazim and thiabendazole) of *Ascochyta pisi* and *Ascochyta pinodes* in the laboratory after reports of decreased effectiveness in the field in France during 1990 and 1991. However, there have been no reports of resistance to thiabendazole in the UK (Biddle, 1994) and it is still recommended for *Ascochyta* spp. control where it is current practice to treat seed when infection is between 5% and 35%.

Microdochium nivale has long been associated with cereal seedling losses in temperate regions (Cristani, 1992). It has recently been associated with yield losses in Ireland (Humphreys *et al.*, 1995) and in trials in France (Sequin and Maumene, 1994). Differences in the efficacy of seed treatments have been reported in trials in Scotland (Burgess *et al.*, 1996; Cockerell, 1997).

A survey of certification authorities by Rennie and Cockerell (1994) found that there was a general agreement that *M. nivale* was of little importance on barley. In the current study however, both Ireland and Germany have listed *M. nivale* as an important pathogen on barley although neither country was able to comment on its economic significance in terms of yield. In Scotland, Richardson *et al.* (1976) found that spring barley seed infected with high levels of *M. nivale* did not appear to show a reduction in germination, but seed infection was associated with higher levels of seedling disease if the seed was sown untreated.

An examination of barley seed samples in south-western Germany in 1992–1994 showed that *Fusarium sporotrichioides* was the most important seedborne pathogen of barley in this area (N. Leist, personal communication). *F. sporotrichioides* is a widespread soil fungus causing seedling or storage rots. It is most significant as a producer of mycotoxins in stored grain (Seemüller, 1968). Although *F. sporotrichioides* has been reported as one of the commonest *Fusarium* spp. on wheat seed in Canada (Duthie *et al.*, 1986) there are few reports of *F. sporotrichioides* found in Western Europe at this time.

The incidence of *Claviceps purpurea* (ergot) in rye and triticale in Bavaria, Germany, increased during 1994 and 1995 (A. Obst, personal communication). A similar increase was recorded in England and Wales in 1994 following adverse weather at flowering (Paveley *et al.*, 1996). The incidence of *C. purpurea* in cereal seed is controlled through seed certification and *C. purpurea* is the only seedborne pathogen with specified maximum levels allowed in seed samples by EC Directives.

Yarham and Jones (1992) reported *T. caries* in wheat crops in the south-east of England during the 1980s. They associated its incidence with an increase in the area sown with untreated farm-saved seed and the ability of the fungus to survive as spores in dry soil between crops. Their results showed that seed treatment could not be relied upon to eliminate the disease which could therefore build up in a seed stock after a few years of sowing untreated farm-saved seed. In Scotland, Oxley and Cockerell (1996) reported that differences exist in the efficacy of seed treatment now available in the UK where seed is contaminated at a level of 1000 spores per seed. In Germany the occurrence of *T. caries* in organically produced wheat has increased in the last 3 years (N. Leist, personal communication). Approximately 40% of UK certified and farm-saved winter wheat seed is contaminated with *T. caries* spores, albeit at very low levels (Cockerell and Rennie, 1996). Where seed is intended to be sown untreated as a means of producing organic wheat, or to reduce costs, it should be tested for *T. caries* contamination.

Sphacelotheca reiliana (head smut of maize) has been a particular problem in south-west France since 1983 and French surveys in 1991 have picked up isolated foci in more northerly regions. The main source of infection is the soil where spores can remain viable for 5 years or more (Decoin, 1992; Imgraben, 1995). However, one of the most important sources of spread is infected seed and this is the most likely explanation for its arrival in a new area. The pathogen can affect both cob and inflorescence. Cob infection is perhaps most serious because, instead of grains, smut spores are formed, resulting in yield losses of between 7 and 15 d t ha^{-1} where up to 20% of heads are infected (Imgraben, 1995). Control measures are available and in France the use of resistant varieties is advocated for the future. At present control is achieved mainly through seed treatment and, where soil is contaminated, seed treatment combined with a microgranular formulation of fungicide applied to furrows (Naibo and Delalondre, 1992). More recently *S. reiliana* has been found in Germany; to prevent spread of the disease in southern Baden farmers were offered flutriafol-treated seed and a special licence was issued to allow the application of the granular formulation of the fungicide for infected areas in 1994 and 1995. Control is of vital importance, particularly in highly susceptible varieties where seed treatment does not give complete control.

Barley seed infected with *Pyrenophora teres* is an important means of introducing the pathogen to new areas (Piening, 1968; Locke *et al.*, 1981). Other sources of inoculum, such as debris, stubble and volunteers are probably more significant in terms of disease development particularly in autumn-sown crops (Jordan, 1981). Disease development at later stages of crop growth is more likely to affect yield in both winter and spring barley (Jordan *et al.*, 1985; Deadman and Cooke, 1987). Yield losses of up to 40% can best be prevented by well-timed field fungicide sprays (Brandl and Hoffmann, 1990; Hims and Gladders, 1994). Seed treatment is an important control measure, particularly in areas where *P. teres* has not been a problem.

Diaporthe helianthi of sunflower caused serious losses in France in 1985 and by 1994 Delos and Moinard (1995) reported that practically every sunflower-growing region of France had become infected due to the spread of aerial inoculum. They concluded that *D. helianthi* was introduced to France from an infected seed source. This is despite the fact that these authors believe the probability of the disease being transmitted by seed is very slight unless the seed becomes polluted with overwintering pycnidia from harvest debris. Careful ploughing of infected plant residues into the soil prevents overwintering of the pathogen (Vörös *et al.*, 1983). Eradication of volunteers in fallow land is essential. Seed treatment can control the seedborne phase and the selection of highly resistant varieties for the future is advocated.

Plasmopara halstedii (sunflower mildew) was recorded in central France during the late 1980s. This was unusual as all French varieties were described as being resistant to race 1 of mildew which was the only race reputed to be present in France (Lafon *et al.*, 1996). Race A appeared in 1988 and race B in 1989. Today a large area of France is affected by *P. halstedii.* A further two races, races C and D, were discovered in 1995. Fortunately, all sunflower varieties that are resistant to races 1, A and B are also resistant to races C and D. Legislation was introduced in 1993 and compulsory seed treatment, with an approved product, of all sunflower varieties not recognized as being resistant to the races of mildew present in France was enforced. In addition, seed coming from third world countries and the European Community where mildew occurs are subject to treatment. More recently legislation has been introduced whereby an approved product must be used to treat all non-resistant varieties. There is no seed test available to determine seed infection and laboratory tests are used to verify the effectiveness of approved seed treatments. Like *D. helianthi* and *S. reiliana*, the soilborne phase of the pathogen is important; however, to prevent spread of the disease the use of appropriate seed treatments and resistant varieties is necessary.

Seed Testing

In recent years seed testing has played an important role in the control and monitoring of important seedborne diseases within Western Europe. However, there are a number of instances where, although vital, the seed test procedure is thought to be inadequate (Table 1.2). Many of the tests deemed inadequate are for autumn-sown cereals where the time between harvest and sowing is very short. In most countries more than 95% of cereal seed is routinely treated and at present there is no requirement in Western European countries to reduce seed treatment inputs. Until suitable seed tests exist there will be no incentive to move towards a practice of treating seed according to need (as in

Table 1.2. Methods available for detection of new and priority seedborne pathogens in Western Europe.[a]

Pathogen	Host	Is seed testing vital?	Are seed test procedures adequate?
Category A			
Ascochyta spp.	pea	yes	no – too expensive
Fusarium sporotrichioides	barley	yes	yes
Microdochium nivale	wheat	no, where routine seed treatment exists with effective fungicides	yes, unless quick results are required to make seed treatment decisions
Pyrenophora graminea	barley	yes	no, duration of test too long and too expensive
Pyrenophora teres	barley	no	yes
Tilletia caries	wheat	no, where routine seed treatment exists with effective fungicides yes, for organic wheat production	no, repeatability of test, duration and cost
(*Fusarium* spp.)	(peas)	(yes)	(yes)
(*Phoma* spp.)	(linseed)	(no)	(no)
Category B			
Claviceps purpurea	rye and triticale	no	yes
Microdochium nivale	barley	yes	no – time consuming and expensive
Microdochium nivale	oats	no	yes
Sphacelotheca reiliana	maize	yes	yes
Plasmopara halstedii	sunflower	yes	no, no method available
(*Diaporthe helianthi*)	(sunflower)	(no)	no, test is time consuming and not standardized

[a]Additional pathogens listed by the seed industry as important but not listed by the pathologists are given in brackets.

some of the Scandinavian countries). Where cereal seed is to be sown untreated it should be tested for seedborne pathogens (see Chapter 14).

The cost of seed testing must provide cost benefits to the growers whether in terms of seed treatment decisions or as part of quarantine or certification procedures. A move toward new technologies, e.g. the polymerase chain reaction (PCR) and DNA probes, could be cheaper in the long term if a number of seed lots could be tested simultaneously and if there was sufficient demand for the tests to cover development costs. Development of this technology is extremely expensive and will only be considered where the benefits over existing methodology are likely to be high. Where seed is moved from one region to another the repeatability of tests within a laboratory and between laboratories is extremely important. Entering seed health methods into the International Seed Testing Association Rules and organizing referee tests would appear to be an essential step in today's culture of quality assurance.

CONCLUSION

This paper highlights a number of seedborne pathogens considered as new or priority diseases in Western Europe. They are all fungi affecting agricultural crops. However, there may be other pathogen–host combinations that could be of equal importance in vegetable and horticultural crops and these might include bacterial and viral pathogens.

Changes in climatic conditions, cultural practices or the use of more susceptible varieties can increase the inoculum levels of some pathogens (*C. purpurea, P. teres, Phoma* spp., *Fusarium* spp., *M. nivale*) very quickly. In individual crops these may cause economic losses. The use of resistant or less susceptible varieties coupled with the appropriate seed treatment is probably the most likely method of control in these cases. However, for many crops breeding for resistance to seedborne fungi may not be considered a priority.

Monitoring pathogen levels and/or screening pesticides for resistance may prevent losses that occur when pathogens become resistant to individual pesticides (*P. graminea, M. nivale, Ascochyta* spp.). This would be expensive but where a single active ingredient is used to control a specific pathogen the cost of monitoring may provide savings in the long term. Alternatively agrochemical companies must continually produce new pesticides with different activity against specific fungi, or a selection of active ingredients must be available to reduce the selection pressure on these pathogens. Where possible the use of heavily infected seed, as determined by seed health tests, should be avoided.

The use of imported seed as a vehicle to introduce new pathogens to areas is still an important consideration for some crops in Western Europe, e.g. sunflower and maize. As shown in the examples given in this chapter, once the pathogen is established, and particularly where a soilborne phase exists, eradication is very difficult. It is essential to test seed moving from areas where serious diseases are known to exist. A reliable seed test must of course be available. The value of such testing would be determined by the economic effect of controlling the new pathogen. The cost of producing new sunflower varieties resistant to *P. halstedii* and *D. helianthi* is very high as the production of new varieties can take 10 years or more.

Large-scale testing of any pathogen for any reason, quarantine, monitoring or seed treatment thresholds, requires sensitive, reproducible, quick and low-cost testing. The cost of providing this is high whether in terms of research into new technologies or the provision of skilled resources. Government policies, the consumer and the seed industry will drive the need to change existing seed testing capabilities in Western Europe.

ACKNOWLEDGEMENTS

I am grateful to the seed pathologists and seed industry representatives who took time to complete my questionnaire.

REFERENCES

Biddle, A.J. (1994) Seed treatment usage on peas and beans in the UK. In: Martin, T. (ed.) *Seed Treatment: Progress and Prospects*. BCPC Monograph no. 57, British Crop Protection Council Publications, Bracknell, pp. 143–149.

Brandl, F. and Hoffmann, G.M. (1990) Untersuchungen zum Befallsverlauf der Netzfleckenkrankheit an Wintergerste (Erreger: *Drechslera teres* (Sacc.) Shoem.; Perfekstadium *Pyrenophora teres* Drechs.) und Erntscheidungshilfen für Fungizidmassnahmen *Gesunde-pflanzen* 42, 192–202.

Burgess, P.J., Fornier, N. and Matthews, S. (1996) Effect of fungicidal seed treatments on the emergence and yield of winter wheat infected with *Microdochium nivale*. In: *Proceedings Crop Protection in Northern Britain, Dundee*, Vol. 1, pp. 85–90.

Cailliez, B. (1994) *Helminthosporium gramineum* de l'orge bien maitrisée. *Cultivar (Paris)* Suppl. 9, 366 pp.

Cockerell, V. (1997) Review of work carried out at the Official Seed Testing Station (OSTS) on the control of *Microdochium nivale* (*Fusarium nivale*) on winter wheat in Scotland (in press). *Proceedings of ISTA Pre-Congress Seminar Copenhagen*, June 1995 (in press).

Cockerell, V. and Rennie, W.J. (1996) *Survey of Seed-borne Pathogens in Certified and Farm-saved Cereal Seed in Britain Between 1992 and 1994*. Report no. 124. Home Grown Cereals Authority Project, Hamlyn House, Highgate Hill, London.

Cockerell, V., Rennie, W.J. and Jacks, M. (1995) Incidence and control of barley leaf stripe (*Pyrenophora graminea*) in Scottish barley during the period 1987–1992. *Plant Pathology* 44, 655–661.

Cristani, C. (1992) Seed borne *Microdochium nivale* (Ces. ex Sacc.) Samuels (= *Fusarium nivale* (Fr.) Ces.) in naturally infected seeds of wheat and triticale in Italy. *Seed Science and Technology* 20, 603–617.

Deadman, M.L. and Cooke, B.M. (1987) Effects of net blotch on growth and yield of spring barley. *Annals of Applied Biology* 110, 33–42.

Decoin, M. (1992) Lutte contre le charbon des inflorescences du mais. Les moyens qui passent par la semence. *Phytoma* 437, 59–62.

Delos, M. and Moinard, J. (1995) Evolution du *Phomopsis* du tournesol en France. Un bref historique. *Phytoma* 473, 22–24.

Diamond, H., Cooke, B.M. and Donne, B. (1995) Information note: *Microdochium* leaf blotch of oats – a new foliar disease caused by *Microdochium nivale* var. *nivale* in Ireland. *Plant Varieties and Seeds* 8, 171–174.

Duthie, J.A., Hall, R. and Asselin, A.V. (1986) *Fusarium* species from seed of winter wheat in Eastern Canada. *Canadian Journal of Plant Pathology* 8, 282–288.

Hims, M.J. and Gladders, P. (1994) Use of fungicide-manipulated epidemics to determine critical periods for disease control and yield response in winter barley. In: *Brighton Crop Protection Conference – Pests and Diseases*, Vol. 1, pp. 283–288.

Humphreys, J., Cooke, B.M. and Storey, T. (1995) Effects of seed-borne *Microdochium nivale* on establishment and grain yield of winter sown wheat. *Plant Varieties and Seeds* 8, 107–117.

Imgraben, H. (1995) Maiskopfbrand – ine neve Krankheit aus Frankreich. *PSP – Pflanzenschutz-Prakis* 2, 36–38.

Jones, D.R., Slade, M.D. and Briks, K.A. (1989) Resistance to organo-mercury in *Pyrenophora graminea*. *Plant Pathology* 38, 509–513.

Jordan, V.W.L. (1981) Aetiology of barley net blotch caused by *Pyrenophora teres* and some effects on yield. *Plant Pathology* 30, 77–87.

Jordan, V.W.L., Best, G.R. and Allen, E.C. (1985) Effects of *Pyrenophora teres* on dry matter production and yield components of winter barley. *Plant Pathology* 34, 200–206.

Lafon, S., Penaud, A., Walser, P., De Guenin, M., Molinero, V., Mestres, R. and Tourvieille, D. (1996) Le mildiou du tournesol toujours sous surveillance. *Phytoma* 484, 35–36.

Locke, T., Thorpe, I.G., Carter, M.A. and Gay, C.E. (1981) Differential sensitivity of *Pyrenophora teres* to organomercury formulations. *Plant Pathology* 30, 89–93.

Molinero, V., Leroux, P., Tivoli, B., Champion, R. and Said, J. (1993) Resistance to benzimidazole fungicides in pathogens of *Ascochyta* diseases in peas. *Seed Science and Technology* 21, 531–535.

Naibo, B. and Delalondre, C. (1992) Charbon des inflorescences du mais. Un concept nouveau par traitement du sol. *Phytoma* 438, 40–41.

Oxley, S. and Cockerell, V. (1996) Transmission of bunt (*Tilletia caries*) in winter wheat with reference to seed-borne inoculum, site and climate. In: *Proceedings Crop Protection in Northern Britain, Dundee*, Vol. 1, pp. 99–102.

Paveley, N.D., Rennie, W.J., Reeves, J.C., Wray, M.W. Slawson, D.D., Clark, W.S., Cockerell, V. and Mitchell, A.G. (1996) *Cereal Seed Health and Seed Treatment Strategies*. Report no. 34. Home Grown Cereals Authority Review, Hamlyn House, Highgate Hill, London, 131 pp.

Piening, L. (1968) Development of barley net blotch from infested straw and seed. *Canadian Journal of Plant Science* 48, 623–625.

Rennie, W.J. and Cockerell, V. (1994) Strategies for cereal seed treatment. In: Martin, T. (ed.) *Seed Treatment: Progress and Prospects*. BCPC Monograph no. 57, British Crop Protection Council Publications, Bracknell, pp. 17–26.

Richardson, M.J., Whittle, A.M. and Jacks, M. (1976) Yield–loss relationships in cereals. *Plant Pathology* 25, 21–30.

Seemüller, E. (1968) Untersuchungen über die morphologische und biologische Differenzierung in der *Fusarium*. Sektion *Sporotrichiella*. In: Smith, I.M., Dunez, J., Lelliott, R.A., Phillips, D.H. and Archer, S.A. (eds) *European Handbook of Plant Diseases*. Blackwell Scientific Publications, p. 280.

Sequin, B. and Maumene, C. (1994) Les risques dus aux maladies transmises par les semences de cereales à paille. In: *ANPP Quatrième Conférence Internationale sur les Maladies des Plantes*. Bordeaux. pp. 557–565.

Vörös, J., Lérànth, J. and Vajna, L. (1983) Overwintering of *Diaporthe helianthi*, a new destructive pathogen on sunflowers in Hungary. In: Smith, I.M., Dunez, J., Lelliott, R.A., Phillips, D.H. and Archer, S.A. (eds) *European Handbook of Plant Diseases*. Blackwell Scientific Publications, p. 315.

Yarham, D.J. and Jones D.R. (1992) The forgotten diseases: why we should remember them. In: *Proceedings Brighton Crop Protection Conference: Pests & Diseases*, Vol. 3, pp. 1117–1126.

Recent Problems with Loose Smut in Oats and Common Bunt in Wheat in the Nordic Countries

2

G. Brodal[1], H. Kortema[2], C. Scheel[3] and K. Sperlingsson[4]

[1]*Norwegian Agricultural Inspection Service, Seed Testing Station, Fellesbygget, N-1432 Ås, Norway;* [2]*PPIC Seed Testing Department, PO Box 111, FIN-32201 Loimaa, Finland;* [3]*Danish Plant Directorate, Skovbrynet 20, DK-2800 Lyngby, Denmark;* [4]*Swedish Seed Testing and Certification Institute, Onsjövägen, S-268 81 Svalöv, Sweden*

INTRODUCTION

Loose smut (*Ustilago avenae* (Pers.) Rostrup) of oats and common bunt (*Tilletia caries* (DC) Tul.) of wheat have been rare or absent in the Nordic countries since the end of the 1950s, when organomercury seed treatments became widely used. However, since about 1990, these diseases have been reported more frequently; in particular the incidence of loose smut of oats is increasing. This paper gives a survey of these diseases as reported from field plots and disease surveys in the Nordic countries from the earliest part of this century through to 1995. At present, a Nordic seed pathology group is standardizing a laboratory method for testing seed for these diseases and assessing the relationship between infection levels detected in laboratory tests and the incidence of infected plants in the field. Preliminary results from this work are discussed.

LOOSE SMUT OF OATS

Observations from Field Plot Samples and Field Inspection

From the 1920s onwards, loose smut of oats has been recorded during the routine seed health testing of field control plot samples at the seed testing stations in the Nordic countries. Occurrence of the disease in Denmark, Norway, Finland and Sweden, recorded as the percentage of infected control plot samples, is presented in Figs 2.1–2.4.

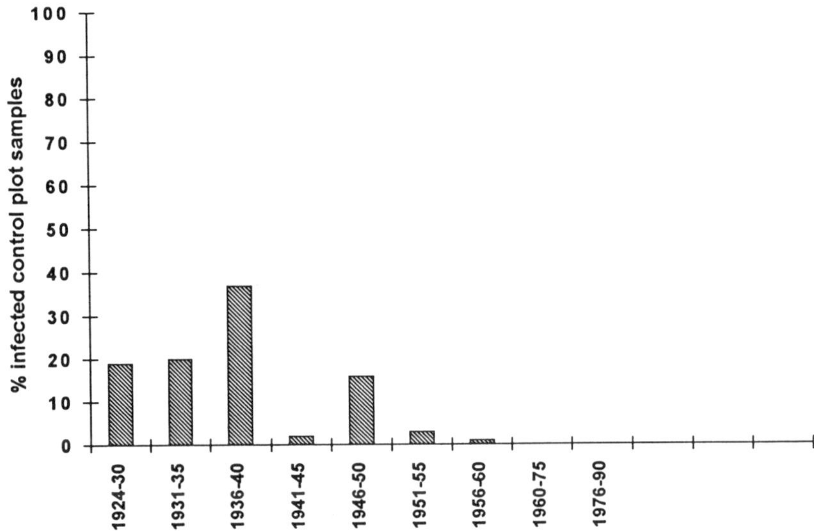

Fig. 2.1. Occurrence of loose smut infected control plot samples of oat seed in Denmark 1924–1990 (based on data from Jørgensen, 1996).

In the first part of this century loose smut of oats occurred quite frequently in all the Nordic countries, but from the beginning of the 1960s, the disease was almost eradicated. In Denmark, the disease has not been recorded since 1957 (Fig. 2.1). However, in Norway the disease was observed in 1989, for the first time in 30 years, in certified seed (Fig. 2.2). In most samples from recent years, the frequency of seed infection was rather low, often less than 0.1%. However, in a few cases, 1–3% (and even up to 10%) infected plants have been observed in control plots. Over the last 30 years in Finland, loose smut of oats has not been completely eradicated and in 1995 a significant increase in the incidence of the disease was observed. About 25% of oat seed lots tested were infected (Fig. 2.3). In Sweden, the occurrence of the disease increased during the 1990s, particularly in 1995, and was widespread with about 40% of the samples tested showing symptoms of infection (Fig. 2.4). Loose smut has also been recorded in Sweden during inspection of seed crops. In 1995, the disease was observed in nearly 60% of the inspected fields, a tremendous increase compared with previous years (Fig. 2.5). In addition, the infection frequency in each field was also significantly increased.

Laboratory Analyses

The occurrence of loose smut in oat seed samples can be tested using the method described by Kietreiber (1984) for common bunt of wheat. However, the method is more difficult to carry out for *U. avenae*, because the spores are considerably smaller than those of *Tilletia caries* and are accordingly more difficult to recognize. One limitation of the method is that only spores and not dormant mycelium can be detected. Infection of *U. avenae* is known to be carried both as mycelium in the pericarp and the inner walls of the glumes and

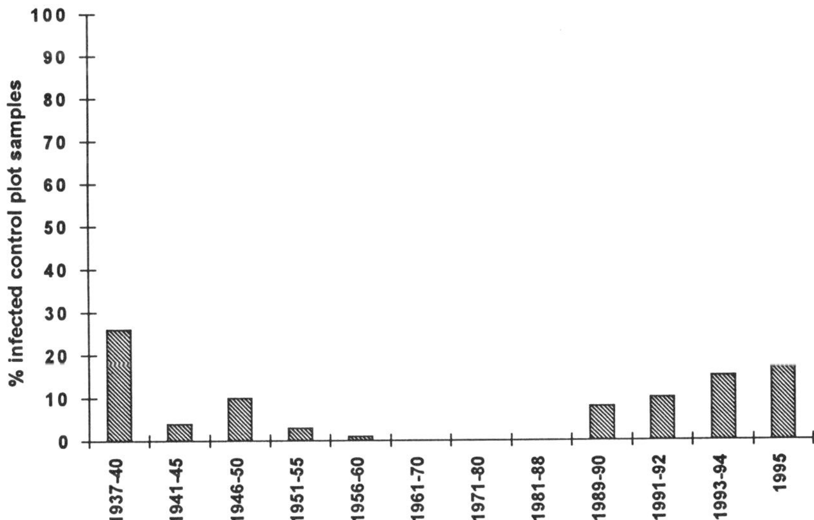

Fig. 2.2. Occurrence of loose smut infected control plot samples of oat seed in Norway 1937–1995.

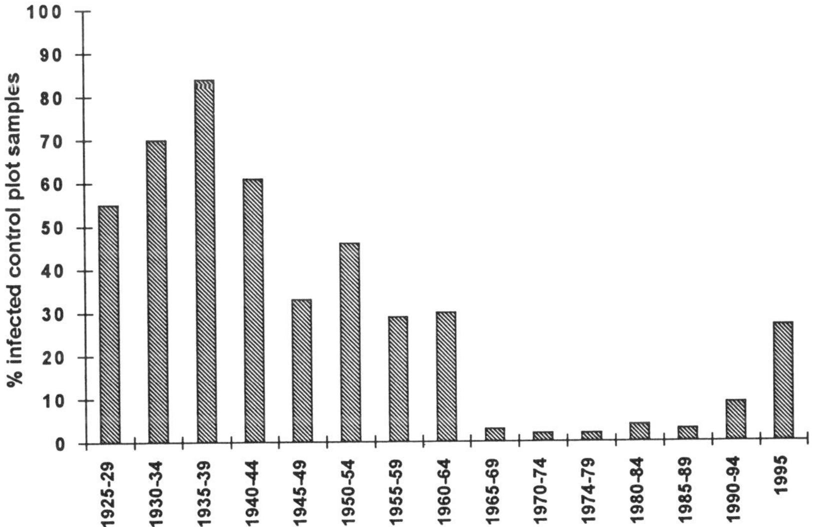

Fig. 2.3. Occurrence of loose smut infected control plot samples of oat seed in Finland 1925–1995.

as spores on the surface of the glumes. Assuming that an infected seed sample will contain spores on the surface of seeds in addition to mycelium in the pericarp, this method will identify all infected seeds lots. In 1995, a Nordic

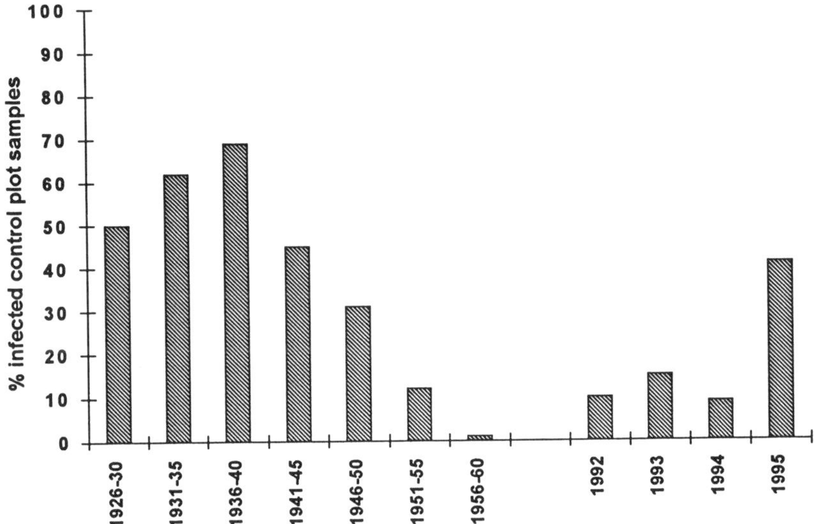

Fig. 2.4. Occurrence of loose smut infected control plot samples of oat seed in Sweden 1926–1995.

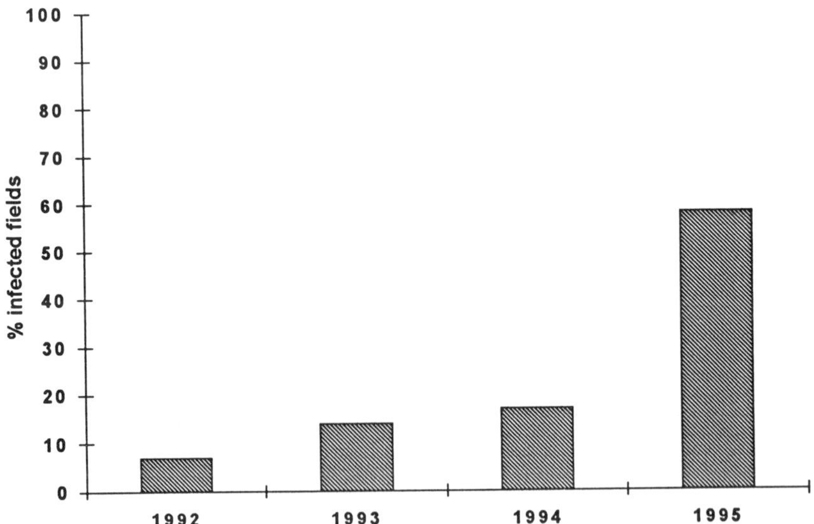

Fig. 2.5. Occurrence of loose smut infection in oat seed production fields in Sweden 1992–1995.

Table 2.1. Number of *U. avenae* spores g^{-1} seed detected in three naturally infected oat seed samples tested in three Nordic laboratories with the 'washing/haemocytometer' method.

	Laboratory number		
Sample number	1	2	3
1	7,900,000	1,200,000	7,100,000
	6,100,000	1,300,000	7,800,000
2	1000	778	1255
	970	3056	1022
3	106,000	85,500	65,050
	26,400	104,500	58,800

working group tested three samples, each consisting of two unknown replicates. Some agreement was found both within and between laboratories, but there is a need for further development and standardization of the method (Table 2.1). In Finland, Norway and Sweden several seed lots have been tested on request from seed companies, to assess the need for treatment. Both clean and infected seed lots have been found. However, the question is: how much infection can be tolerated before treatment is required?

Control Strategies and Tolerance Levels

Beyond the seed certification regulations for smut diseases, which allow no more than 0.05% infected plants in control plots for basic seed and 0.1% infected plants for certified first and second generation, no special precautions have been taken in Norway and Finland to restrict or control the disease. In Sweden, however, the situation was considered so serious after the 1995 season, that treatment with an effective chemical was compulsory for all basic seed of oats, unless a laboratory test showed that treatment was unnecessary. Treatment of basic seed was demanded if more than 100 spores g^{-1} seed were detected, in cultivars known to be susceptible, and at 200 spores or more in less susceptible cultivars (Swedish Seed Testing and Certification Institute, 1995). Seed treatment of certified seed first and second generation was not demanded but recommended if the laboratory test revealed more than 200 and 500 spores g^{-1} seed, respectively, for the two categories of susceptibility.

To establish the relationship between seed and field infection, the Seed Testing Station in Norway made a preliminary comparison of data from laboratory tests and infection frequencies seen in field plots (Table 2.2). The results obtained were variable. Some samples showed that a low incidence of seed infection resulted in a low infection level in the field plots, but other samples with a similar level of infection produced more diseased plants. In addition to the somewhat unreliable laboratory procedure used, a possible explanation is that seed infection can sometimes exist as mycelium within the seed in the absence of spores. In this event the method would be inappropriate, suggesting that an improved method is required.

Example of Yield Losses from Loose Smut of Oats

An example of the yield losses that loose smut of oats can cause is shown in Table 2.3. In two seed lots of oats, naturally infected with loose smut, a treatment with carboxin + imazalil (Fungazil C) reduced the incidence of infected plants on average from about 25% to 0.2% for the two lots used, increasing the yield by about 30%.

COMMON BUNT OF WHEAT

Disease Surveys

The severity of attack of common bunt (*Tilletia caries*) in Denmark from 1905 to 1995 is shown in Fig. 2.6. The disease was rare and of only minor importance when mercury seed treatments were used routinely (Nielsen and Jørgensen, 1994). However, from 1989 onwards, attacks of common bunt of winter

Table 2.2. Occurrence of *U. avenae* in oat seed samples tested by laboratory analyses and field plot growing at the seed testing station in Norway 1994 and 1995.

Sample number	Number of spores g^{-1} seed	% Infected plants
1	111	0.01
2	583	0.01
3	6600	0.05
4	9100	0.10
5	89	0.11
6	400	0.80
7	34,900	1.23
8	53,200	1.50
9	89,300	3.30

Table 2.3. Occurrence of loose smut in a field trial with untreated and treated naturally infected seed lots of oats in Norway 1995.

Seed lot	Seed treatment	Number of spores g^{-1} seed	% Infected plants	Yield (relative)
1	untreated	1,800,000	34.0	100
1	imazalil + carboxin	–	0.2	138
2	untreated	10,000,000	17.0	100
2	imazalil + carboxin	–	0.2	127
2	imazalil + guazatine	–	5.6	111

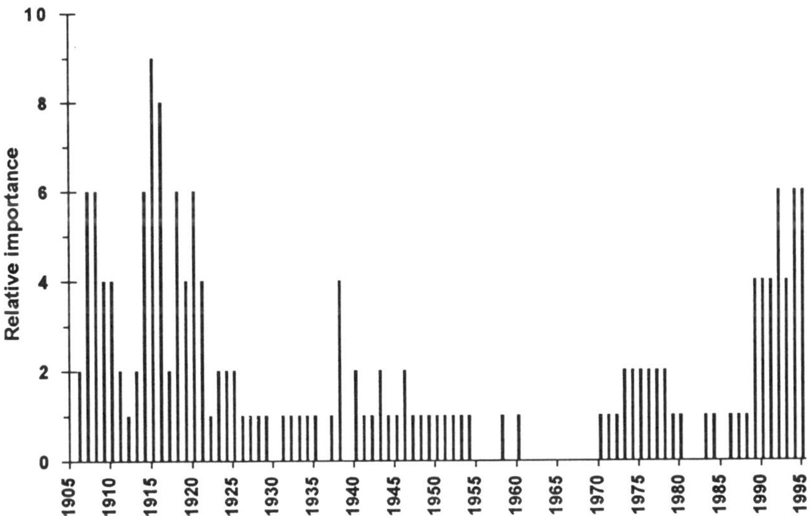

Fig. 2.6. The severity of attack of common bunt in Denmark, estimated from disease surveys made by the Danish Institute of Plant and Soil Science (from Stapel *et al.*, 1976; Stapel and Nielsen, 1992; unpublished survey report, Danish Institute of Plant and Soil Science).

wheat were reported more frequently. In Sweden, Johnsson (1991) concluded that the quantity of *T. caries* spores detected in Swedish winter wheat was low and had been for many years. However, a severe attack of common bunt was reported in winter wheat in 1994 (Johnsson *et al.*, 1995). The increased occurrence of this disease in Norway in recent years is illustrated by the increase in the numbers of inquiries to the Seed Testing Station and the Plant Protection Institute from grain merchants about the feeding quality of smutted grain.

Laboratory Analyses

Laboratory analyses, using the washing/haemocytometer method, of wheat seeds for *T. caries* spores in Denmark commenced in 1989. Samples were for both certification and advisory purposes. Infections were recorded in approximately 10–50% of the samples tested. However, in most cases, fewer than 100 spores g^{-1} seed were detected. There has been no significant increase in the number of infected samples since 1989. In Sweden, a laboratory test is compulsory for all winter wheat, and not more than 500 spores g^{-1} seed are accepted. However, most seed lots are found to be clean, as are the few samples tested each year in Finland and Norway. In the Nordic countries if *T. caries* infection is detected then seed treatment is always recommended.

To develop a standardized procedure for testing of wheat seeds for common bunt the Nordic countries working group has organized comparative tests with

Table 2.4. Number of *T. caries* spores g^{-1} seed detected in four naturally infected wheat seed samples tested in three Nordic laboratories with the 'washing/haemocytometer' method.

	Laboratory number		
Sample number	1	2	3
1	11	0	2333
	20	0	867
2	1,500,000	277,500	199,650
	1,000,000	268,500	220,200
3	1,000,000	85,650	126,700
	1,000,000	121,500	89,400
4	1000	1111	689
	370	360	3577

Table 2.5. Common bunt on two cultivars inoculated with different number of spores (from Hansen, 1976).

	% Infected plants	
Number of spores g^{-1} seed	Cultivar 1	Cultivar 2
0	0	0
6000	0.6	3.5
25,000	0.7	10.9
100,000	12.3	27.8
400,000	34.0	20.3
1,600,000	77.3	63.2

the washing/haemocytometer method. As with the test for loose smut of oats, the results from this trial were not satisfactory (Table 2.4). These data can partly be explained by each laboratory introducing a slight variation in the way they carried out the test. Attempts to standardize the method further are in progress.

Example of Inoculum–disease Relationship

Table 2.5 shows an example of the relationship between the quantity of seed-borne inoculum and the frequency of infected plants in field plots (Hansen, 1976). Even the lowest amount of inoculum gave significant infection.

DISCUSSION

The increased occurrence of loose smut of oats and common bunt of wheat is in part believed to be related to the replacement of mercury compounds with less efficient products, and/or reduced fungicide seed treatment. In the 1980s, mercury seed treatments were gradually replaced by other products and from around 1990 mercury was prohibited in Nordic countries (Brodal *et al.*, 1994). In addition to mercury being a broad-spectrum, effective fungicide, it was also a very cheap insurance against seedborne diseases. Because the most effective fungicides today are rather expensive, some farmers may choose to use cheaper and less effective chemicals, or use untreated seeds. However, the prohibition of mercury is not considered to be the only reason for the recurrence of the diseases. In Norway, it was estimated that during the period 1960–1990, approximately 50% of oat seed had been treated with mercury, and some of the seed lots in which the disease recurred had been treated with a mercury compound. No comparison has been made of the susceptibility of new and old cultivars to the loose smut pathogen. However, the disease appears to be more widespread in some cultivars than others (Sperlingsson, 1996) and is partly limited to cultivars arising from the same initial breeding material. There is reason to believe that some of the new cultivars are more susceptible to loose smut than older cultivars grown 20 years ago. Such cultivar differences in susceptibility to loose smut are well known in North America (Wilcoxson and Stuthman, 1993).

Although wheat seed infection with *T. caries* seem to be rare and mostly controlled by seed treatment, increased occurrence of the pathogen has been reported in all Nordic countries. This increase is partly believed to be caused by the use of untreated seed or with treatment with a less effective fungicide. In Denmark, it has been decided only the most effective products will be approved (Nielsen and Jørgensen, 1994) and correct application of the seed treatment will be emphasized. In addition, the role of soilborne inoculum, as a consequence of reduced soil tillage, cannot be excluded, especially where wheat is sown after wheat (Johnsson *et al.*, 1995). Also the recent increase in the winter wheat area might contribute to a higher occurrence of the disease.

To accomplish the aim of only treating cereal seeds according to need (Brodal, 1993), the main challenge facing the Nordic countries will be to work against the recent problems with loose smut of oats and common bunt of wheat. First of all, it will be necessary to produce healthy seed, through strict standards in the seed regulations. Treatment of seed with effective fungicides will be an important action. But for both environmental and economical reasons, treatment of seed should be based on a rational assessment. To be able to assess the need for seed treatment, we need to improve the existing testing methods, and to develop a method for detection of *U. avenae* mycelium in oat seed. Further, it is important to collect information about the inoculum–disease relationship to establish thresholds for recommendation for fungicide treatment and for rejection of seed lots. Development of resistant cultivars should also be given high priority.

ACKNOWLEDGEMENTS

We are grateful to Signe Dahl for her assistance with preparation of tables and figures. This study was financially supported by the Nordic Council of Ministers.

REFERENCES

Brodal, G. (1993) Fungicide treatment according to need in the Nordic countries. In: *Crop Protection in Northern Britain*, The Association for Crop Protection in Northern Britain 1993, pp. 7–16.

Brodal, G., Halkilahti, A.M., Jørgensen, J. and Sperlingsson, K. (1994) [Seed borne plant diseases and seed pathological investigations.] Frøbårne plantesjukdommer og frøpatologiske undersøkelser. TemaNord 630, Nordisk Ministerråd, København, 50 pp.

Hansen, L.R. (1976) *Bunt (Tilletia caries) on Wheat in South-East Norway in 1973 and 1974.* Report no. 68. Norwegian Plant Protection Institute, Ås, pp. 633–643.

Johnsson, L. (1991) Survey of common bunt (*Tilletia caries* (DC) Tul.) in winter wheat during the period 1967–1987 in Sweden. *Journal of Plant Diseases and Protection* 98, 67–72.

Johnsson, L., Hedene, K.A. and Gustafsson, G. (1995) Common bunt and dwarf bunt in winter wheat in Sweden in 1994. In: *36th Swedish Crop Protection Conference, Agriculture.* Swedish University of Agricultural Sciences, Uppsala, pp. 205–218.

Jørgensen, J. (1996) *Plant Pathological Records in Annual Reports of the Danish State Seed Testing Station 1871–1990.* Johs. Jørgensen, Måløv, Denmark, 72 pp.

Kietreiber, M. (1984) Bunt (stinking smut). *ISTA Handbook on Seed Health Testing.* Working Sheet no. 53. International Seed Testing Association, Zürich, 4 pp.

Nielsen, B.J. and Jørgensen, L.N. (1994) Control of common bunt (*Tilletia caries* (DC) Tull.) in Denmark. In: Martin, T. (ed.) *Seed Treatment: Progress and Prospects*, pp. 47–52. British Crop Protection Council, Monograph no. 57.

Sperlingsson, K. (1996) Loose smut of oats – history and present situation. In: *37th Swedish Crop Protection Conference, Agriculture.* Swedish University of Agricultural Sciences, Uppsala, pp. 233–241.

Stapel, C., Jørgensen, J. and Hermansen, J.E. (1976) Sædekornets sygdomme i Danmark, deres udbredelse, betydning og bekæmpelse ved afsvampning, især i perioden 1906–1975. *Tidsskrift for Landøkonomi* 163, 185–283.

Swedish Seed Testing and Certification Institute (1995) Betning av vårspannmål. *Information 1995: 4*, 1 p.

Wilcoxson, R.D. and Stuthman, D.D. (1993) Evaluation of oats for resistance to loose smut. *Plant Disease* 77, 818–821.

Seed Health Testing in Eastern and Central European Countries – the Present and Prospects

3

K. Tylkowska

Department of Seed Science and Technology, Poznań Agricultural University, Baranowo, 62-081 Przeźmierowo, Poland

INTRODUCTION

Seed health has been considered an attribute of high quality and is of great concern both to the seed industry and growers. To avoid unnecessary seed treatment and application of fungicide to the growing crop, which is costly and sometimes harmful to the environment, sowing healthy seed is a priority. In recognition of this, seed health testing has been included as an integrated part of seed quality control systems in many countries. This paper reviews the current practice and prospects of seed health testing in several Eastern and Central European (former COMECON) countries and is based mostly on responses to a questionnaire distributed among seed inspection services and research institutes.

RANGE OF SEED HEALTH TESTING

Cereal seeds (barley, *Hordeum vulgare* L.; rye, *Secale cereale* L.; triticale, × *Triticosecale* Wittm.; and wheat, *Triticum aestivum* L.) are viewed as the highest priority amongst agricultural crops in this region. Table 3.1 shows selected seedborne pathogens from a range of these agricultural crops. The tolerance levels for seed infection by these pathogens are reported.

Attention has been paid almost exclusively to the presence of fungi in seed of various crops. According to the available information, *Xanthomonas campestris* pv. *phaseoli* (Smith) Dye in bean seeds is the only bacterial pathogen

 Seed Health Testing (eds J.D. Hutchins and J.C. Reeves)

Table 3.1. Selected seedborne pathogens to be detected and tolerance levels for their occurrence.

Host	Pathogen	Tolerance level (%, unless otherwise stated)	Country
Hordeum vulgare	*Drechslera* spp.	2.0	Czech Republic
	U. nuda	0.3 pre-basic and basic	Belarus[c]
	Ustilago sp.	0.2 breeder, pre-basic and basic, 0.5 certified	Russia[d]
Linum usitatissimum	*C. linicola*	3.0	Hungary
	F. oxysporum f.sp. *lini*	2.0	Belarus[c]
	P. lini	0	Hungary
	S. linicola	0	Hungary
Pisum sativum	*Ascochyta* spp.	3.0 basic, 7.0 certified	Czech Republic
	A. pisi	8.0	Slovakia
	Ascochyta spp. and/or *B. cinerea* and/or *Fusarium* spp.	8.0[a]	Poland
	P. herbarum and *Fusarium* spp.	5.0 basic, 7.0 certified	Czech Republic
Triticum aestivum	*Fusarium* spp.	3.0	Czech Republic
	S. nodorum	10.0	Czech Republic
	T. tritici	15 spores per seed breeder, pre-basic and basic 100 spores per seed certified	Russia[d]
	U. tritici	0.2 breeder, pre-basic and basic, 0.5 certified	Russia[d]
Vicia faba	*A. fabae*	3.0 basic, 7.0 certified	Czech Republic
	A. fabae and/or *B. cinerea* and/or *B. fabae* and/or *Fusarium* spp. and/or *S. sclerotiorum*	5.0 pre-basic and basic[b]	Poland
	Fusarium spp.	5.0 basic, 8.0 certified	Czech Republic

[a]The sum of infested seeds of *P. sativum.*
[b]The sum of infested seeds of *V. faba.*
[c]1996–1998.
[d]Fyodorova, 1995

for which restrictions exist (the Czech Republic) and no tests for viruses have been included so far in routine seed health analyses. Depending on the need, sometimes on request, some seed lots are tested for the presence of additional seedborne fungi. This applies to cereals, including maize, grasses, rapeseed, sugarbeet and various leguminous crops. Results of seed health tests are published regularly in the Czech Republic and Slovakia or occasionally in Hungary in annual reports.

Further research on the relationships between seed infestation levels and disease development under natural conditions is required. To minimize the risk of disease, threshold levels must be developed for the environmental conditions in which seed is sown. Afterwards, necessary changes in legislation should be taken into consideration. However, before coming into the national standards, determination of the pathogens could be performed on an advisory basis, on request of a seed producer or seed buyer.

Most attention has been paid to agricultural and some vegetable crops. Forest tree seed health management in the Czech Republic deserves special recognition. This is an interesting example of how recommendations for storage or treatment are given according to seed health status, in terms of the class of seed infestation (Anon., 1972; Procházková, 1996). Should our attention be directed exclusively towards pathogens in seed lots designated for sowing? Since it is well recognized that many pathogens can survive for an extremely long time, health testing of seeds designated for long-term storage in gene banks should also be taken into consideration.

METHODS FOR DETECTION OF SEEDBORNE PATHOGENS

Compared with the other seed quality tests, seed health testing requires more technically advanced equipment, media, reagents, etc. Laboratory personnel also require frequent and intensive training. Until now, seed health analysts have usually been trained by specialists from research institutes, agricultural universities and plant protection services.

Methods used for detection of seedborne pathogens often differ from one country to another. Table 3.2 details how different methods are used to detect seedborne pathogens in a range of different countries. International Seed Testing Association (ISTA) methods are not always universally adopted (e.g. for testing for *Ascochyta* spp. and *Colletotrichum lindemuthianum*). However, to compare results obtained for the same pathogen on an international level, the methods used should be standardized and must give results that are accurate, reliable and reproducible. This could be achieved through the development of methodology by Central and Eastern European laboratories conducted under the framework of the ISTA Plant Disease Committee (PDC). All respondents to the questionnaire indicated strong interest in such a proposal. The host–pathogen combinations of the highest priority requiring investigation depend very much on the country concerned and priorities must be set.

Development of research and routine work in seed pathology requires more recognition and financial support. Establishment of one or two well equipped laboratories where seed health tests and research work could be performed, would appear to offer a reasonable solution for reduction of costs.

Table 3.2. Methods used for the detection of seedborne pathogens.

Pathogen/host	Method	Country
Ascochyta spp./pea	blotter test	Poland
	ISTA	Czech Republic
Ascochyta spp. broad bean	agar test (Czapek–Dox)	Slovakia
	blotter test	Poland
Colletotrichum lindemuthianum/bean	agar test (PDA)	Czech Republic
	ISTA	Hungary, Poland
C. linicola/flax	agar test (PDA)	Czech Republic
	blotter test	Hungary
Drechslera spp./barley	agar test	Slovakia
	ISTA	Czech Republic
Fusarium spp./pea	blotter test	Czech Republic
Fusarium spp./bean, broad bean, pea	blotter test	Poland
Fusarium spp./cereals	agar test/blotter test	Czech Republic
	agar test	Slovakia
Tilletia spp./cereals	dry seed inspection (purity test)	Poland
	washing test (ISTA?)	Slovakia, Belarus

PLANT AND SEED HEALTH MANAGEMENT

Field Inspection

Field inspection plays an important role in seed certification schemes and production of high quality seed. Attention should be directed towards field inspection for symptoms of seedborne disease. Links between field inspection services and seed health testing laboratories are required. In the Czech Republic, a report on the plant health status from a field inspection serves as a source of information for the seed testing laboratory. In Slovakia, wheat crops are inspected for *Tilletia controversa*, where a nil infection standard is enforced. In both Hungary and Poland there are lists of pathogens which are examined for in a field inspection and for which tolerance levels are set.

Strategies for Seed Treatment

The quantity of seed treatments used could potentially be minimized if treatments were applied when they were required, based on the results of seed health tests. After the withdrawal of mercury-based seed treatments in Central Europe (Poland, 1977; Hungary, 1978; the Czech Republic and Slovakia, 1991; Bulgaria, 1992) a range of fungicides suitable for seed treatment have been registered (Jackson *et al.*, 1994; Pruszyński, 1995). The efficacy of several widely available commercial fungicide products against stinking smut in Polish wheat has been reported (Kubiak and Korbas, 1996).

Apart from Hungary, where seed treatment against *Tilletia* spp. has been obligatory for 25 years, the recommendation for treatment is, at least in part, based upon the results of seed health tests. In Slovakia quantities of fungicide active ingredients applied to seeds are often monitored by colorimetric methods. In Hungary, seed health tests are repeated after seed treatment to test the efficacy of the treatment. In Poland, vegetable seeds are retested after chemical treatment to monitor their capacity for germination. Similarly, in the Czech Republic germination tests may be performed on some treated seed lots.

CHANGES IN THE OCCURRENCE OF PATHOGENS

Development of many diseases depends on environmental, especially climatic conditions. Generally the occurrence and severity of seedborne diseases varies from year to year and from region to region according to weather conditions during the growth of the crop and/or seed harvest. Cultural practices, including effective seed treatment and fungicidal sprays in the field also affect disease development. In Slovakia, a gradual decrease in wheat seed infection with *T. controversa* was observed between 1976 and 1993, most probably due to appropriate seed and crop management. In contrast, in Poland, due to a reduction in the use of seed treatments (partly for economic reasons), the occurrence of stinking smut in wheat has increased in the last two seasons. Measures have now been taken to tackle this problem: a premium will be paid to growers who have their seed properly treated with an appropriate seed dressing.

It is well known that diseases may also be introduced into new, previously unaffected areas as a result of intensive seed trade. There have been cases in the Czech Republic where an increase in bacterial diseases has been observed, and in Poland, where the occurrence of several viruses was noticed for the first time in 1995. Specifically these include cucumber green mottle mosaic tobamovirus and sunn hump mosaic tobamovirus (Pospieszny *et al.*, 1996).

COOPERATION

Various aspects of seed pathology have been investigated in universities and research institutes in all the countries of the region. A closer coordinated approach between research groups and seed inspection services is necessary to solve problems arising within the field of seed pathology. Besides cooperation within a country, there are several examples of international collaboration, e.g. between the Czech Republic and Austria, the Czech Republic and Slovakia, and Hungary and The Netherlands. This approach must be developed further.

At present there seems to be insufficient dissemination of epidemiological data, methodology and information related to seedborne diseases and their control. This should be rectified in the near future. Attention must be directed towards further publicising work developed by the ISTA-PDC (including Working Sheets and *Handbook on Seed Health Testing*) and must incorporate the involvement of Central and Eastern European laboratories in method development, meetings and training courses. Interest in such courses with particular regional emphasis, covering techniques used for the detection of priority seed-borne pathogens in Central and Eastern Europe, has been expressed by many of the respondents. A meeting with these objectives was organized under the auspices of the ISTA-PDC in 1985 (Tylkowska and Grzelak, 1996).

ACKNOWLEDGEMENTS

I greatly appreciate the help of S.F. Buga (Belarus), M. Houba, J. Chod, B. Kokošková, J. Krátká and Z. Procházková (Czech Republic), P. Békési and M. Jakab-Kondor (Hungary), B. Michalik, J. Dąbrowska, C. Jańczak, M. Korbas and colleagues from seed testing stations from Poland, A. Vitáriušová and M. Tátarová (Slovakia), whose active cooperation made this work possible. My thanks are also due to C.J. Langerak for his critical review of the manuscript.

REFERENCES

Anonymous (1972) *Quality Tests of Forest Tree Fruits and Seeds*. CSN 48 1211. Prague, 35 pp.

Fyodorova, R.N. (1995) Recommendations on permissible levels of seed-borne infection of wheat and barley in Russia. Poster and Seed Symposium Abstracts. XXIV ISTA Congress, Copenhagen, June 7–16, p.61.

Jackson, D., Marshal, R., Tomkins, M.J., Novy, O., Papp, Z. and Romaniuk, J. (1994) The role of carboxin + thiram FS in post-mercury seed treatment strategies in Central Europe. In: Martin, T. (ed.), *Seed Treatment: Progress and Prospects*. BCPC Monograph no. 57, pp. 123–128.

Kubiak, K. and Korbas, M. (1996) Możliwość zwalczania śnieci cuchnącej pszenicy. [Possibility of controlling stinking smut disease in wheat]. Poster at the XXXVI Sesję Naukową Instytutu Ochrony Roślin, Poznań, 8–9 February 1996.

Pospieszny, H., Zielińska, L. and Cajzą, M. (1996) Nowe wirusy roślinne w Polsce. [The new plant viruses in Poland]. In: *Tezy do referatów i komunikatów na XXXVI Sesję Naukową Instytutu Ochrony Roślin*, Poznań, 8–9 February 1996, pp. 41–42.

Procházková, Z. (1996) Health testing of forest tree seeds in the Czech Republic. In: Mathur, S.B. and Mortensen, C.N. (eds) *Proceedings of the ISTA Pre-Congress Seminar on Seed Pathology 'Seed Health Testing in the Production of Quality Seeds'*, Copenhagen, 6 June 1995, pp. 81–97.

Pruszyński, S. (ed.) (1995) *Zalecenia ochrony roślin na lata 1995/1996 dotyczące zwalczania chorób, szkodników oraz chwastów roślin uprawnych. Cz. I i II.* Instytut Ochrony Roślin, Poznań.

Tylkowska, K. and Grzelak, K. (1996) Seed pathology and seed health testing in Poland. In: Mathur S.B. and Mortensen, C.N. (eds) *Proceedings of the ISTA Pre-Congress Seminar on Seed Pathology 'Seed Health Testing in the Production of Quality Seeds'*, Copenhagen, 6 June 1995, pp. 39–45.

The Recurrence of *Burkolderia solanacearum* in Southern European Countries

4

M. Sousa Santos

Direcção Geral de Protecção das Culturas, Tapada da Ajuda, Ed. 1, 1300 Lisboa, Portugal

INTRODUCTION

Pseudomonas solanacearum E.F. Smith, the causal agent of potato brown rot, also known as bacterial wilt disease (Kelman, 1953; Hayward, 1994), was renamed *Burkolderia solanacearum* (Smith) Yabuuchi *et al.* (1992) and recently proposed to the genus *Ralstonia* (Yabuuchi *et al.*, 1995). It is a quarantine organism included in the A2 List of EPPO (EPPO, 1990), as well as in the Annex I-A2 of the European Union Plant Health Directive 95/4/EC (European Union, 1995). Its host range is one of the widest of all phytopathogenic bacteria. The most susceptible plant family, in number of species affected, is the *Solanaceae* (Kelman, 1953). Susceptible species are also found in more than 50 other plant families. The number of new host species continues to increase, ranking bacterial wilt as one of the most important plant diseases of bacterial origin in the world (Kelman *et al.*, 1994).

B. solanacearum has mainly been reported in tropical, subtropical and warm temperate regions of the world. In Europe, it was first detected at the beginning of the century but its occurrence was sporadic. Immediate control measures were taken and the pathogen was eradicated. Recurrences in 1989 and 1995 in the European Union led to the affected countries establishing strict control measures aimed at eradicating the pathogen (EPPO, 1996).

SUBSPECIFIC CLASSIFICATION

The bacterium is a very heterogeneous species which does not have a uniform biology and host range. It is a complex of variants traditionally described as groups, races, biovars, biotypes, subraces and strains. These different classifications have caused confusion in the literature. Buddenhagen *et al.* (1962) divided the species into three races on the basis of host range. Race 1 attacks potato, tomato, tobacco, eggplant and other *Solanaceae*, diploid banana and several herbaceous plants, its optimal temperature being 32–36°C. Race 2 is pathogenic to triploid banana, *Heliconia* and other tropical *Musaceae*, occurring mostly in Central and South America. Race 3, the so called 'potato race', affects mostly potato, tomato and eggplant, with an optimal temperature range of 25–28°C and potentially considered as the most dangerous for the temperate zones. Recently, two other races were described affecting *Zingiber* (race 4) and *Morus* (race 5, Buddenhagen, 1986).

Five biovars have been distinguished according to the ability to utilize and/or oxidize three hexose alcohols and three disaccharides. On addition of other phenotypic criteria the apparent homogeneity within biovars may diminish, as has been shown for different biovar 2 isolates (Hayward, 1994). Except for race 3 strains (potato race), which are always in biovar 2 (Hayward, 1994), the correlation between biovar classification and race differentiation is not perfect (Buddenhagen, 1986). Strains belonging to one race do not necessarily belong to the same biovar. Cook and Sequeira (1994) have distinguished several restriction fragment length polymorphism (RFLP) groups within the species, which, essentially, correspond to the genomic fingerprint types described by Gillings and Fahy (1994), forming two genetically distinct divisions with origins in Australasia and the Americas.

INOCULUM SOURCES AND DISSEMINATION

B. solanacearum may survive in soil between successive plantings of a susceptible crop, in plant debris and latently infected potato tubers (Graham and Lloyd, 1978). Other infected or infested vegetable plant material (Hayward, 1991), the rhizosphere of weed hosts (Granada and Sequeira, 1983) and the deeper soil layers (Graham and Lloyd, 1979) may also contribute to its survival. The range and variety of weed hosts are extensive (Kelman, 1953), their significance varying greatly in different environments and cropping systems (Hayward, 1991), some of them being symptomless carriers (Ciampi and Sequeira, 1980). Most of these studies have been performed in tropical areas.

The pathogen has also been detected on perennial and symptomless plants of *Solanum dulcamara* in Sweden (Olsson, 1976) and the UK (Walker, 1992), as well as in *S. cinereum* in a cool temperate region of Australia (Graham and

Lloyd, 1978), indicating the possibility of these weeds acting as inoculum sources for the long-term survival of the pathogen. However, complete or reliable information concerning many aspects of the biology and ecology of the pathogen in temperate areas is still missing, thus showing the urgent need for thorough studies to fill this gap.

True seeds have been reported as a mean of survival and dissemination of the pathogen (Kelman *et al.*, 1994) There are, however, conflicting reports on the occurrence of seed transmission, possibly as a consequence of the low numbers of bacteria present on the seed surface and the consequent difficulty in their detection by conventional methods. Such a source of inoculum, although not prevalent, could have an epidemiological significance in some situations (Prior *et al.*, 1994).

An epiphytic phase in the life cycle of the bacterium has also been suggested as a survival mechanism (Hayward, 1991). Waste and drainage and wash-water from potatoes with brown rot may also constitute an important inoculum source (Olsson, 1976). High populations of the pathogen in surface soil near infected plants after rainfall and in surface furrow waters also suggests an above-ground source of inoculum (Ono, 1983). Injured roots, stem wounds and stomata may provide a means of entry into the plant. The importance of root-knot nematodes (*Meloidogyne* spp.) providing points of ingress into plants is already established (Hayward, 1991). The synergistic interactions between these nematodes and *B. solanacearum* on a variety of hosts is widely recognized (Kelman, 1953; Reddy *et al.*, 1979).

RECENT DETECTIONS IN THE EUROPEAN UNION

After the first detection in Belgium in 1989–1991 near the Belgian/Dutch border further reports were made in the UK in 1992, Belgium in 1993–1994 and The Netherlands in 1993. In all cases, detection occurred in ware-potato fields and the source of inoculum was thought to be associated with the use of contaminated irrigation water or was from an undetected source (Olsson, 1976; Walker, 1992; Caffier and Hervé, 1996; EPPO, 1996). In the UK, potatoes had not been produced in the affected fields for 5–7 years (Walker, 1992). The pathogen was subsequently isolated from *S. dulcamara* weeds present at the water edges of the River Thames (EPPO, 1996). The possible association of the inoculum source with sewage effluents from potato processing plants should not be disregarded (Digat and Caffier, 1996). In all cases, severe control measures were taken and the pathogen was considered to be eradicated.

In 1994, the French Plant Protection Services noticed the first cases of contaminated tomato and eggplant grown under glasshouse in soilless culture in two regions, away from the potato production areas. Drastic control and eradication measures were taken, but the origin of the inoculum source was not unequivocally determined (Caffier and Hervé, 1996). The following year, the disease recurred on tomato in the same location, as well as in a new open-air field. At the same time and far from the traditional potato-growing areas, another disease outbreak was detected on potato. In this case, the seed potato used was of French origin and known to be free from *B. solanacearum* and so it was thought the source of inoculum was the river water used for irrigation (Caffier and Hervé, 1996). In the same year, detection in Portugal during April (Sousa Santos and Henriques, 1995) and Italy in June (Mazzucchi, 1995) aggravated the situation. This time, the inoculum source was determined in both countries and irrefutably linked to Dutch certified seed potato. Research carried out in The Netherlands after those occurrences unequivocally showed the presence of the bacterium in several Dutch ware and seed-potato fields (EPPO, 1996).

SITUATION IN PORTUGAL AND ACTIONS TAKEN

The first symptoms of potato brown rot were detected in April 1995 by the plant protection authority from the Portuguese Regional Services of 'Beira Litoral'. The disease was found in six small ware-potato fields in different locations in the region of Cantanhede, belonging to the same farmer. In all these fields, a single certified seed lot, cv. Jaerla, had been used for planting. The soil in these fields was light and sandy and, when irrigated, the water came from deep wells. The pathogen was promptly identified as *B. solanacearum* race 3, biovar 2 at the laboratory of the 'Direcção Geral de Protecção das Culturas' (DGPC) and action taken to obtain information about the quantity of this seed lot used and to determine the extent of the outbreak. Potato tuber samples from all the potato fields located in the vicinity of the infested fields were tested and shown to be free from the pathogen. The seed lots used in these fields were either from other cultivars or from different lots of cv. Jaerla, confirming the suspicion that the inoculum source was one specific certified seed lot. Meanwhile, all plants from infected fields were eliminated with a herbicide and all vegetative material deep buried under quarantine conditions.

The plant protection authorities from all the Portuguese Regional Services were informed of the situation. Inspections on potato fields started immediately, potato samples with and without symptoms being collected and tested at the DGPC laboratory. Furthermore, television programmes and an information brochure were issued by DGPC (Henriques *et al.*, 1995). As a consequence of these actions, 25 new infected ware-potato fields (in three regions of the country) were detected which had been seeded with seed potato from the same lot or from other lots of the same cultivar, or from lots of two other cultivars (Kondor and Desirée). The presence of the pathogen on all the 'presumably' infested fields (originating from the infected seed lots) could not always be confirmed because of the difficulty in tracing the farmers that used the infected seed and identifying infected crops after the crop had been harvested. These fields are now under close surveillance.

To contain and eradicate the pathogen, the Portuguese Plant Protection Services defined the following compulsory actions to be taken for fields where the pathogen had been detected:

- Destruction of the aerial part of potato plants (by application of herbicide) and the production (by approved disinfection and disposal or deep burial under strict surveillance of the Regional Plant Health authority).
- Quarantine observation of infested fields for 5 years.
- Restriction of cropping practices, specifically to include no solanaceous crops.
- Destruction of all solanaceous weeds and volunteers from infested farms.
- Prevention of irrigation water running from infested to adjacent fields and avoidance of cultural practices conducive to increases in soil pH.
- Implementation of hygiene and access measures to prevent movement of soil (via machinery, vehicles, livestock and personnel) and disinfection of all equipment that has been in contact with the infected seed lot.
- Sampling and testing at the DGPC laboratory of plants and tuber samples from all fields in the neighbourhood of the infested fields and the corresponding stocks.
- Immediate reporting of the occurrence of suspect symptoms of brown rot.

For those fields considered as 'potentially' infested, a close surveillance was established and the same containment and eradication measures were taken, when appropriate. Further actions taken by the DGPC laboratory include developing rapid and sensitive detection and identification methods, so that reliable results may be obtained when plant material and complex substrates are tested.

Besides those measures, additional actions taken in 1996 consist of:

- Testing potato seed lots coming from the European Union (EU) or from third countries with 'equivalence' status as well as all ware-potato consignments imported from non-EU countries.
- Analysis of soil, irrigation water and volunteers from infested, 'potentially' infested and adjacent fields, to monitor the presence of the pathogen.
- Inspection and testing of all seed-potato fields before issuing a plant passport.
- Country-wide inspection of ware-potato crops and laboratory analysis at DGPC of tuber samples collected at random during the inspections, with special reference to those areas located in the vicinity of the infested fields.
- Inspection of tomato fields in areas where the pathogen was detected.
- Continuing the national surveys of *B. solanacearum*.

CONCLUSIONS AND PROSPECTS

Bacterial wilt has continued to be a significant problem throughout the world despite all the efforts developed by many national and international organizations. Biovar 2 (race 3) of the pathogen is very damaging to solanaceous crops including potato, tomato and eggplant in all areas of the world where it occurs. In Europe many aspects of the biology and epidemiology of the pathogen are still unknown, in contrast with the profuse literature on these subjects from tropical and subtropical areas (Hayward, 1991). The variability of the pathogen, its distribution and its means of survival and dissemination are unknown. Some information is available about latent survival in infected crops and weeds (Ciampi and Sequeira, 1980) and possible spread over long distances through latently infected tubers, true botanical seed (Hayward, 1991) and irrigation water (Olsson, 1976). However, it is not known if the currently available detection methods are sensitive enough to detect very low levels of the pathogen generally associated with latent infections, particularly when testing complex substrates which often contain inhibitory compounds.

Despite its status as quarantine organism for the EU (European Union, 1995), recent outbreaks in Europe have shown that this pathogen represents a serious threat under both northern and southern European climatic conditions. Potential yield losses under European conditions are currently unknown but in France complete losses in potato, tomato and eggplant have occurred. A similar situation occurred in Portugese and Italian potato crops (Sousa Santos and Henriques, 1995; Digat and Caffier, 1996). Should the pathogen become established in the EU, very serious implications could arise for movement of seed- and ware-potato within or between member states, as well as for exports of seed to non-EU countries. In addition, prohibition of the growth of solanaceous crops for several years following outbreaks of bacterial wilt will have devastating consequences on commercial growers. A heavy burden will also be placed on plant health services, due to the required increase in field inspection, sampling and laboratory analysis.

All the EU Plant Health authorities are increasing efforts to prevent further dissemination within the Union as well as to eradicate the pathogen from infested fields, so that confidence in the quality of EU seed- and ware-potato may be re-established. Intensive and integrated research cooperation between EU experts on quarantine bacteria is also required, not only to standardize detection protocols with appropriate sensitivity and specificity but also to provide the necessary epidemiological information on the behaviour and distribution of the pathogen within Europe and to assess the risks involved to production and export. In order to standardize diagnostic protocols for obligatory use in EU quarantine laboratories for implementation of the Council Directive for *B. solanacearum* (awaiting approval), a group of experts working under the Plant Health Committee has set up an interim test scheme for detection and identification of this pathogen.

ACKNOWLEDGEMENTS

Gratitude is expressed to Dr David Caffier for sending information from papers at proof stage and to Dr Nigel Lyons for helpful comments and review of the manuscript.

REFERENCES

Buddenhagen, I.W. (1986) Bacterial wilt revisited. In: Persley G.J. (ed.) *Bacterial Wilt Diseases in Asia and South Pacific*. ACIAR Proceedings Vol. 13, ACIAR, Canberra, Australia, pp. 126–143.

Buddenhagen, I.W., Sequeira, L. and Kelman, A. (1962) Designation of races in *Pseudomonas solanacearum*. *Phytopathology* 52, 726 (abstract).

Caffier, D. and Hervé, A. (1996) La pourriture brune de la pomme de terre, un nouveau risque bactérien en Europe. *La Pomme de Terre Française* 492, 20–23.

Ciampi, L. and Sequeira, L. (1980) Influence of temperature on virulence of race 3 strain of *Pseudomonas solanacearum*. *American Potato Journal* 57, 307–317.

Cook, D. and Sequeira, L. (1994) Strain differentiation of *Pseudomonas solanacearum* by molecular genetic methods. In: Hayward, A.C. and Hartman, G.L. (eds), *Bacterial Wilt: The Disease and its Causative Agent, Pseudomonas solanacearum*. CAB International, Wallingford, pp. 77–93.

Digat, B. and Caffier, D. (1996) Apparition d'une redoutable maladie en France: le fletrissement bactérien des solanées. *Phytoma* 482, 33–37.

EPPO (1990) Quarantine procedure no. 26. *Pseudomonas solanacearum*, inspection and test methods. *EPPO Bulletin* 20, 255–262.

EPPO (1996) *Situation of Burkolderia (Pseudomonas) solanacearum in the EPPO Region*. EPPO Reporting Service 1996, 96/002.

European Union (1995) Commission Directive 95/4/EC amendment of 21 February 1995 to the European Community Plant Health Directive (77/93/EEC). *Official Journal of the European Communities* L44, 28.02.95, 56–60.

Gillings, M. and Fahy, P. (1994) Genomic fingerprinting: toward a unified view of the *Pseudomonas solanacearum* species complex. In: Hayward, A.C. and Hartman, G.L. (eds) *Bacterial Wilt: The Disease and its Causative Agent, Pseudomonas solanacearum*. CAB International, Wallingford, pp. 95–112.

Graham, J. and Lloyd, A.B. (1978) An improved indicator plant method for the detection of *Pseudomonas solanacearum* race 3 in soil. *Plant Disease Reporter* 62, 35–37.

Graham, J. and Lloyd, A.B. (1979) Survival of potato strain (race 3) of *Pseudomonas solanacearum* in the deeper soil layers. *Australian Journal of Agricultural Research* 30, 489–496.

Granada, G.A. and Sequeira, L. (1983) Survival of *Pseudomonas solanacearum* in soil, rhizosphere and plant roots. *Canadian Journal of Microbiology* 29, 433–445.

Hayward, A.C. (1991) Biology and epidemiology of bacterial wilt caused by *Pseudomonas solanacearum*. *Annual Review of Phytopathology* 29, 65–87.

Hayward, A.C. (1994) Systematics and phylogeny of *Pseudomonas solanacearum* and related bacteria. In Hayward, A.C. and Hartman, G.L. (eds), *Bacterial Wilt: The*

Disease and its Causative Agent, Pseudomonas solanacearum. CAB International, Wallingford, pp. 123–136.

Henriques, L., Carvalho, S.S. and Carrinho, H. (1995) *Doença do Pús da Batateira – Pseudomonas solanacearum*. Instituto de Protecção da Produção Agro-Alimentar, Lisbon, 6 pp.

Kelman, A. (1953) *The bacterial wilt caused by Pseudomonas solanacearum*. North Carolina Agricultural Experiment Station Technical Bulletin 99, 194 pp.

Kelman, A., Hartman, G.L. and Hayward, A.C. (1994) Introduction. In Hayward, A.C. and Hartman, G.L. (eds), *Bacterial Wilt: The Disease and its Causative Agent, Pseudomonas solanacearum*. CAB International, Wallingford, pp. 1–7.

Mazzucchi, U. (1995) Avvizzimento batterico: pericolosa malattia per la patata in Italia. *L'Informatore Agrario* 31, 65–68.

Olsson, K. (1976) Experience of brown rot caused by *Pseudomonas solanacearum* (Smith) Smith in Sweden. *EPPO Bulletin* 6, 199–207.

Ono, K. (1983) Ecological studies on the bacterial wilt of tobacco caused by *Pseudomonas solanacearum* E.F. Smith III. Distribution and spread of the pathogen in infected tobacco field under rainfall. *Bulletin of the Okayama Tobacco Experiment Station* 42, 149–153.

Prior, P., Grimault, V. and Smith, J. (1994) Resistance to bacterial wilt (*Pseudomonas solanacearum*) in tomato: present status and prospects. In: Hayward, A.C and Hartman, G.L. (eds) *Bacterial Wilt: The Disease and its Causative Agent, Pseudomonas solanacearum*. CAB International, Wallingford, pp. 209–223.

Reddy, P.P., Singh, D.B. and Kishun, R. (1979) Effect of root-knot nematode on the susceptibility of Pusa Purple Cluster Brinjal to bacterial wilt. *Current Science* 48, 915–916.

Sousa Santos, M. and Henriques, L. (1995) *Identificação de Pseudomonas solanacearum (Smith) Smith em Batateiras Provenientes de Campos de Produção de Batata de Consumo do Concelho de Cantanhede*. Centro Nacional de Protecção da Produção Agrícola, CPA/D-1, PPA(ID)-38/95, CNPPA, Lisbon, 17 pp.

Walker, D. (1992) *Potato Brown Rot, Pseudomonas solanacearum*. Central Science Laboratory Plant Disease Notice no. 73, 5 pp.

Yabuuchi, E., Kosako,Y., Oyaizu, H., Yano, I., Hotta, H., Hashimoto, Y., Ezaky, T. and Arakawa, M. (1992) Proposal of *Burkolderia* gen. nov. and transfer of seven species of the genus *Pseudomonas* homology group II to the new genus, with the type species *Burkolderia cepacia* (Palleroni and Holmes 1981) comb. nov. *Microbiology and Immunology* 36, 1251–1275.

Yabuuchi, E., Kosako, Y., Yano, I., Hotta, H. and Nishiuchi, Y. (1995) Transfer of two *Burkolderia* and *Alkaligenes* species to *Ralstonia* gen. nov. – proposal of *Ralstonia pickettii* (Ralston, Palleroni and Douderof, 1973) comb. nov., *Ralstonia solanacearum* (Smith, 1896) comb nov. and *Ralstonia eutropha* (Davis, 1969) comb. nov. *Microbiology and Immunology* 39, 897–904.

Seedborne Diseases of Economic Importance in Israel and the Use of Alternatives to Mercuric Fungicides

5

N. Lisker[1] and J.D. Klein[2]

[1]*Unit of Seed Pathology and Department of Agronomy and Natural Resources;* [2]*Department of Seed Science and Genetic Resources, Institute of Field and Garden Crops, ARO, The Volcani Center, Bet Dagan 50250, Israel*

INTRODUCTION

In response to the cancellation of the registration of the mercuric fungicides as seed protectants in Israel, studies were initiated to identify alternative seed treatments. The major crops affected by the removal of mercuric fungicides for seeds were wheat, cotton and groundnuts, which together occupy the majority of the area sown to field crops in Israel. These crops are all affected by fungal attack which results in poor stands which in some cases demand replanting (as in the case of cotton damping-off). The organisms involved include the weak parasites *Alternaria*, *Aspergillus niger*, *Rhizopus* in wheat, and *Rhizoctonia solani* in the case of cotton, and *Aspergillus niger* and sometimes *Rhizopus* in groundnut. Most recently, *Fusarium* infestations in basil seeds have begun threatening a major Israeli herb export crop.

WHEAT

One of the major factors contributing to poor stands in wheat fields in Israel seems to be a complex interaction between certain environmental conditions and the presence of weakly parasitic seedborne and/or soilborne fungi. Wheat is sown in Israel at the beginning of November, when rainfall resumes after the summer drought. Seeds germinate in the wet soil, and if rain continues during November and December, seedlings emerge and develop normally. However, if germination is followed by a dry period (a situation that occurs every 3 or 4

 Seed Health Testing (eds J.D. Hutchins and J.C. Reeves)

years), imbibed seeds may stop developing or even dry back. This results in extreme drought stress under which weakly parasitic fungi such as *Alternaria, Rhizopus* and *Aspergillus* may infect weakened seeds. Seedlings may never emerge, even if rain falls later in December. Post-germination drought stress is common, especially on the edge of the Negev desert located in the south of Israel, where wheat is widely cultivated.

Until its registration was cancelled, Caspan (ethoxymercuric chloride) was the only fungicide used for wheat seed treatment in Israel. It was introduced in the beginning of the 1950s as a seed treatment against smuts (*Tilletia caries* and *T. foetida*) (Nevo, 1959) and continued to be used as a prophylactic fungicide against occasional infestation by weak parasites. No reports on the activity of mercuric fungicides against weak parasites or on the role of the micro-organisms on wheat germination were available.

To simulate drought stress in the laboratory, wheat seeds were artificially weakened (Battacharyya *et al.*, 1985; Baskin, 1987; Ram and Wiesner, 1988; Pandey, 1989; Lisker, 1990). Six spring wheat (*Triticum aestivum* L.) cultivars were chosen because of their variation in 1000-kernel weight, plant height, heading date and resistance to diseases (Pnuel, 1988). Visually normal intact seeds were obtained locally and were examined on germination paper, in sterile sand and in clay, silty and sandy silty loam soils.

With normal seeds sown on germination paper, percentage germination ranged from 96% to 100% regardless of fungicide treatments. However, certain fungicides were phytotoxic towards shoot development in different cultivars. Only thiram, oxine-copper, Panoctine and tebuconazole had no phytotoxic effect. Fungi (mostly *Rhizopus* spp.) were observed on less than 30% of the non-germinated seeds. Thus, seedborne fungi did not seem to be a major factor in reducing germination. These fungicides had no significant effect on root development.

When experiments were carried out in soils, the differences in the percentage of emerging seedlings or in seedling development between the treatments were more pronounced. Again, no phytoxicity was observed after treatment with the four non-phytotoxic fungicides mentioned above. Solel *et al.* (1986) also reported no phytotoxicity for thiram treatments in wheat stands under field conditions, although other fungicides investigated have been found to be phytotoxic (Lisker, 1990). It is interesting to note that the percentages of germination and of seedling emergence and development in seeds not treated with fungicides were very high. These results confirm that no phytopathological problems will appear in Israel if intact seeds are not exposed to unfavourable environmental conditions during germination and seedling emergence, as noted previously (Solel *et al.*, 1986). The phytopathological conditions obtained in Israel are thus quite different from those in wheat growing areas in Europe and North America where cool wet conditions frequently lead to significant damage as a result of seedborne pathogens such as *Septoria nodorum, Helminthosporium* spp. and *Fusarium* spp.

The most prevalent fungi in non-treated wheat seed in Israel were *Rhizopus* spp. (Lisker, 1990). When seeds were treated with sodium hypochlorite (NaOCl), *Rhizopus* spp. almost disappeared and *Alternaria* spp. became the dominant fungi (> 95%). *Alternaria* spp. seem to be mainly located deep inside the seeds while *Rhizopus* spp. develop externally. Fungicide treatment initially reduced the percentage of seeds (45%) on which fungal colonies developed. After 7 days, however, in fungicide-treated seeds, there was a considerable increase in the percentage of seeds on which fungal colonies sporulated, demonstrating that the seed treatments exerted only a fungistatic effect. Some of the phytotoxic fungicides caused a significant change in the seed mycobiota. In these cases, *Alternaria*, the dominant fungus in NaOCl-treated seeds, almost disappeared, and either *Aspergillus tamari* or a combination of *Saccharomyces* and *Sporobolomyces* became the dominant fungi.

When artificially weakened seeds were sown in sterile sand, seedling emergence from fungicide-treated seeds was very high (90%), regardless of the seed treatment applied. This emphasizes the role of weak soil fungal pathogens present in natural soils in decreasing seedling emergence. It is possible, however, that under natural field conditions, the weak seedborne parasitic fungi act synergistically with soil fungi in causing poor stands (Lisker, 1990).

Only thiram and oxine-copper caused a consistent increase in the rate of seedling emergence in natural soils. Since thiram was already widely in use in Israel on other crops, it was also applied initially to wheat seeds as a substitute for mercuric fungicides. Although the results obtained were very promising, farmers complained about itching, rashes and skin and respiratory problems, so the use of thiram was discontinued. Currently, oxine-copper is the only wheat seed treatment in use in Israel. In order to avoid a recurrence of the situation with mercuric fungicides when they were the only fungicides in use and their registration was removed, alternative fungicides are currently being evaluated.

COTTON

For many years, mercuric fungicides were used worldwide in cotton to control damping off. In the USA, where mercuric fungicides lost their registration in the early 1970s (Ranney, 1971), there were many fungicide combinations available to cotton growers. However, damping off continues to be a major problem, probably due to the involvement of more than one pathogen (Minton and Garber, 1983; Garber *et al.*, 1989; Watkins, 1981). In contrast, in Israel only *Rhizoctonia solani* was isolated from seedlings showing damping off. Attempts to control the disease have been described (Solel and Palti, 1960; Solel and Minz, 1964; Solel *et al.*, 1986). A survey to determine the magnitude of the disease or the efficacy of mercuric fungicide seed treatments was recently carried out (Lisker and Meiri, 1992).

Damping off seems to be exclusively a soilborne disease that can be contained by seed treatments. Field examinations were performed in Israel during 1988/89, the last season in which mercury-containing compounds were used as seed treatments (Lisker and Meiri, 1992). The fields studied were representative of the majority of cotton-growing areas in Israel and included the cultivars Acala SJ-2 and Pima S-5, as well as some cultivars not yet commercially introduced.

R. solani was the predominant pathogen isolated from most of the diseased seedlings, regardless of the cultivar examined. *Macrophomina phaseolina* was isolated from a few locations, while *Pythium* spp., which are important pathogens in the USA (Minton and Garber, 1983; Garber *et al.*, 1989), were never isolated from diseased cotton seedlings in Israel (Lisker and Meiri, 1992). *R. solani* continues to be the predominant pathogen probably because cotton is mostly grown in monoculture in Israel. In all cases where a high incidence of *R. solani* was observed, the seeds had been treated with mercuric fungicides. In one case, a field was even partially replanted because of the damping off caused by the pathogen. In general, there was a good correlation between field stands and the presence of *R. solani* in diseased seedlings, but in a very few cases there was a relatively high incidence of the pathogen in seedling lesions without a subsequent decrease in field stands. In these cases, the lesions were small and contained inside the hypocotyl tissue. It is possible that the pathogen attacked the seedling at a later stage, when there is enhanced disease resistance (Bateman, 1970).

Since *R. solani* was the sole pathogen associated with poor stands in Israel, the search for alternative fungicides was more successful than in the USA, where a variety of fungi are responsible for poor stand establishment (Garber *et al.*, 1989). Most of the alternatives to mercuric compounds studied for adoption in Israel had already been recognized for the control of *R. solani* in cotton seedlings in other countries (Minton and Garber, 1983): quintozene (PCNB) (Arndt, 1953; Solel *et al.*, 1986); quintozene plus captan (Sinclair, 1958); quintozene plus 11.3% etridiazole, or with carboxin (Ranney, 1971; Huppatz *et al.*, 1983) tolclophos methyl and pencycuron plus captan (Hillocks *et al.*, 1988). In Israel, seed treatments with quintozene plus etridiazole, tolclophos methyl plus thiram, carboxin plus captan or pencycuron plus captan significantly increased the rate of seedling emergence and decreased disease severity in both cultivars, regardless of whether the soil was naturally infested or not. In naturally infested soils, cv. Acala SJ-2 was more resistant to *R. solani* than Pima S-5. Today, quintozene plus etridiazole and tolclophos methyl plus thiram dominate the seed treatment market in Israel. Since the introduction of these treatments 6 years ago, there have been no reports of *R. solani* in Israeli cotton fields, and it is hoped that damping off has been overcome.

With the withdrawal of mercuric fungicide from use in Israel and with the introduction of alternative fungicides as seed treatments, it will be important to monitor the appearance of new pathogens due to fungicide changes. It has already been reported that chemical suppression of one pathogen may increase the activity of another (Bird *et al.*, 1957). As a consequence, in order to broaden the activity spectrum of the fungicides, a mixture of compounds may sometimes be required in order to prevent the build-up of other pathogens.

GROUNDNUT

Pre-emergence damping off in groundnut (*Arachis hypogaea* L.) is caused mainly by *Rhizopus* spp. (Clinton, 1960), which is favoured by high soil moisture and relatively cool weather. Groundnuts are sown in Israel during the onset of summer, when there can be sporadic cool weather. In the past, pre-emergence damping off caused by *Rhizopus* was occasionally a great problem due to the use of seeds extracted from insufficiently dried pods or seeds stored at a high relative humidity before planting. Only with the use of low-moisture kernels for planting did the disease almost disappear from groundnut fields in Israel. Post-emergence damping off, caused by the *Aspergillus niger* group, and known as *Aspergillus* crown rot (Morwood, 1953), develops in warm weather and is the most prevalent disease of groundnut seedlings in Israel (Frank, 1969). The *A. niger* group can be soilborne as well as seedborne (Jackson, 1962), and some members of the group cause post-emergence damping off (Gibson, 1953; Morwood, 1953). This fungus is the dominant fungus in the mycobiota of Israeli groundnut kernels (Joffe and Lisker, 1969; Lisker *et al.*, 1994).

In a survey of groundnut fields in Israel, it was found that 20% of the plants suffered from crown rot during the first 2 weeks after emergence; the rate of additional affected plants decreased thereafter until the onset of flowering (Frank, 1969). From flowering until harvest, wilt of individual stalks or whole plants was found sporadically. Fungi of the *A. niger* group were isolated from diseased seedlings and from the bases of lateral branches and collars of affected adult plants; the disease was identified as *Aspergillus* crown rot. In some cases, significant amounts of *Rhizopus* could also be isolated from diseased seedlings (Frank, 1969).

Mercuric fungicides for groundnut seed treatments were used in Israel until the beginning of the 1960s. The fungicide penetrates into the seed tissues (Jackson, 1962) and eliminates *Rhizopus* spp. from the external parts of the seed. Seed treatment with mercuric fungicides was abandoned in Israel in favour of other safer and more efficient treatments which had less severe side effects (Bell and Jackson, 1968; Frank, 1969). Today, two kinds of fungicides are widely in use for groundnuts in Israel: a mixture of captan and PCNB, or TCMTB (Busan 30) (Cohen *et al.*, 1993). Contradictory results have been observed with these fungicides. TCMTB was initially widely used because of its liquid formulation, which diminishes the dispersal of toxic particles to the environment and/or to farmers. However, it was found that the fungicide can cause certain seedling abnormalities and impair germination. On the other hand, while PCNB controls *Rhizopus* efficiently, captan controls *A. niger* to a large extent but not completely (due to fungal penetration inside the seeds). The problem of *Aspergillus* crown rot in Israel requires further research efforts to provide a completely effective fungicide.

BASIL

A new seedborne disease which is now striking Israel is *Fusarium* wilt and crown rot of sweet basil. Basil (*Ocimum basilicum* L.) is grown as a herb and for medicinal purposes. It is a relatively new crop in Israel, where it is grown commercially in monoculture in heated greenhouses as a major herb crop for export. In recent years, symptoms of *Fusarium* wilt of basil – growth retardation, wilt and occasionally plant death – have been observed. *Fusarium oxysporum* f. sp. *basilici* has been isolated at high frequency from symptomatic plants (Gamliel *et al.*, 1996).

Seedborne *Fusarium* wilt of basil has been reported from several countries, including southern Russia, France, Italy and the USA (Martini and Gullino, 1991; Elmer *et al.*, 1994). Since the types of sweet basil grown in Israel are local selections from European and American cultivars, it is quite possible that the disease arrived in imported seeds. In Israel, the disease has occurred in all basil-growing regions and progresses rapidly during the growing season. In addition to root rot and stem lesions, which are evident at all stage of growth, stems of diseased plants are often covered with a pink–orange layer of macroconidia (Gamliel *et al.*, 1996).

As many as 40% of diseased plants tested in Israel carried seed infested with *F. oxysporum* f. sp. *basilici*. Gamliel *et al.* (1996) also tested the potential of seeds originating from diseased plants to give rise to a new generation of diseased plants and infested seeds by planting the seed in pathogen-free soil. Of the plants that emerged from such seeds, about 50% showed disease symptoms after only 5–9 weeks. More than 50% of the seeds extracted from the second-generation diseased plants were infested with *F. oxysporum* f. sp. *basilici*.

Fusarium is difficult to control with fungicide treatments. Soil sterilization is only partially effective in the control of the disease, since it inhibits the disease only in young plants. Soil sterilization combined with solarization of the greenhouse, irrigation implements and plant support strings is more effective in the control of the disease throughout plant development. The fact that the disease is spread by a great variety of sources and that the pathogen has been detected in air traps in the greenhouse indicates the need for a more effective means of control. An integrated control programme seems most appropriate, starting with pathogen-free seed followed by sterilization of the greenhouse. The development of resistant cultivars could also alleviate the problem.

ACKNOWLEDGEMENTS

We acknowledge the contribution from the Agricultural Research Organization, The Volcanic Center, Bet Dagan 50250, Israel, No. 2036.E, 1996 series.

REFERENCES

Arndt, C.H. (1953) Evaluation of fungicides as protectants of cotton seedlings from infections by *Rhizoctonia solani. Plant Disease Reporter* 37, 397–400.

Baskin, C.C. (1987) Accelerated ageing test. In: Fiala, F. (ed.) *Handbook of Vigour Test Methods.* International Seed Testing Association Publication, Zurich, pp. 43–48.

Bateman, D.F. (1970) Pathogenesis and disease. In: Parmeter, J.R., Jr (ed.) *Rhizoctonia solani, Biology and Pathology.* University of California Press, Berkeley, California, pp. 161–171.

Battacharyya, S., Hazra, A.K. and Sen-Mandi, S. (1985) Accelerated ageing of seeds in hot water: germination characteristics of aged wheat seeds. *Seed Science and Technology* 13, 683–690.

Bell, D.K. and Jackson, C.R. (1968) *Effects of Seed Treatment Fungicides & Seed Quality on the Incidence of Fat Leg of Peanut.* University of Georgia, College of Agriculture Experiment Stations, Research Report 21, 14 pp.

Bird, L.S., Ranney, C.D. and Watkins, G.M. (1957) Evaluation of fungicides, mixed with the covering soil at planting as a control measure for the cotton seedling disease complex. *Plant Disease Reporter* 41, 165–173.

Clinton, P.K.S. (1960) Seed-bed pathogens of groundnuts in the Sudan, and an attempt at control with an artificial testa. *Empire Journal of Experimental Agriculture* 28, 211–222.

Cohen, R., Rivan, Y., Grinstein, A., Kritzman, G. and Halfon-Meiri, A. (1993) Adaptation of a bio-assay technique to test coverage of groundnut seeds with Busan 30. *Hassadeh* 74, 131–133. (In Hebrew with English summary.)

Elmer, W.H., Wick, R.L. and Haviland, P. (1994) Vegetative compatibility among *Fusarium oxysporum* f. sp. *basilicum* isolates recovered from basil seed and infected plants. *Plant Disease* 78, 789—791.

Frank, Z.R. (1969) Localisation of seed-borne inocula and combined control of *Aspergillus* and *Rhizopus* rot of groundnut seedlings by seed treatment. *Israel Journal of Agricultural Research* 19, 109–114.

Gamliel, A., Katan, T., Yunis, H. and Katan, J. (1996) Fusarium wilt and crown rot of sweet basil: involvement of soilborne and airborne inoculum. *Phytopathology* 86, 56–62.

Garber, R.H., Weir, W.L., Paplomatas, E.J., Wakeman, R.J. and DeVay, J.E. (1989) Control of cotton seedling pathogens with chemical and biological seed and soil treatments. In: *Beltwide Cotton Production Conference,* National Cotton Council of America, Memphis, Tennessee, p. 38.

Gibson, J.A.S. (1953) Crown rot, a seedling disease of groundnuts, caused by *Aspergillus niger. Transactions of the British Mycological Society* 36, 198–209.

Hillocks, R.J., Chinodya, R. and Gunner, R. (1988) Evaluation of seed dressing and in-furrow treatments with fungicides for control of seedling disease in cotton caused by *Rhizoctonia solani*. *Crop Protection* 7, 309–313.

Huppatz, J.L., Phillips, J.N. and Wirzens, B. (1983) Laboratory and glasshouse studies of the activity of carboxamide derivatives against *Rhizoctonia solani* in cotton. *Plant Disease* 67, 45–47.

Jackson, C.R. (1962) *Aspergillus* crown rot of peanut in Georgia. *Plant Disease Reporter* 46, 888–892.

Joffe, A.Z. and Lisker, N. (1969) The mycoflora of fresh and subsequently stored groundnut kernels on various soil types. *Israel Journal of Botany* 18, 77–87.

Lisker, N. (1990) Improving wheat seedling emergence by seed-protectant fungicides. *Crop Protection* 9, 439–445.

Lisker, N. and Meiri, A. (1992) Control of *Rhizoctonia solani* damping-off in cotton by seed treatments with fungicides. *Crop Protection* 11, 155–159.

Lisker, N., Michaeli, R. and Frank, Z.R. (1994) *Aspergillus flavus* and other mycoflora of groundnut kernels in Israel and the absence of aflatoxin. *Mycotoxin Research* 10, 47–55.

Martini, P. and Gullino, M.L. (1991) Seed transmission of *Fusarium oxysporum* f. sp. *basilicum* causal agent of basil vascular wilt. *Informatore Fitopatologico* 9, 59–61. (In Italian.)

Minton, E.B. and Garber, R.H. (1983) Controlling the seedling disease complex in cotton. *Plant Disease* 67, 115–118.

Morwood, R.B. (1953) Peanut pre-emergence and crown rot investigations. *Queensland Journal of Agricultural Science* 10, 222–235.

Nevo, D. (1959) Fungicide treatment of cereal seeds – against what? *Hassadeh* 39, 1216–1217. (In Hebrew.)

Pandey, D.K. (1989) Short duration accelerated ageing of French bean seeds in hot water. *Seed Science and Technology* 17, 107–114.

Pnuel, Y. (1988) Sowing of wheat for grain. *Hassadeh* 69, 40–41. (In Hebrew.)

Ram, C. and Wiesner, L.E. (1988) Effects of artificial ageing on physiological and biochemical parameters of seed quality in wheat. *Seed Science and Technology* 16, 579–589.

Ranney, C.D. (1971) Effective substitutes for alkyl mercury seed treatments for cottonseed. *Plant Disease Reporter* 55, 285–288.

Sinclair, J.B. (1958) Greenhouse screening of certain fungicides for control of *Rhizoctonia* damping-off of cotton seedlings. *Plant Disease Reporter* 42, 1084–1088.

Solel, Z. and Minz, G. (1964) Furrow application of fungicides for control of cotton seedling damping-off in Israel. *Indian Phytopathology* 17, 46–50.

Solel, Z. and Palti, J. (1960) Cotton seed rot and damping-off. *Hassadeh* 40, 685–686. (In Hebrew.)

Solel, Z., Meiri, A., Sharir, A., Rimon D., Ziv., O. and Sali, G. (1986) Alternative fungicides for wheat and cotton seed treatments. Report for the years 1985/86. *Gan, Sadeh vaMeshek* August, 15–18. (In Hebrew.)

Watkins, G.M. (1981) *Compendium of Cotton Diseases*. The American Phytopathological Society, St Paul, Minnesota, 87 pp.

New and Important Seedborne Diseases in Southern Africa

6

J.J. Serfontein

Plant Protection Research Institute, Agricultural Research Council, Private Bag X134, Pretoria 0001, South Africa

INTRODUCTION

Seed pathology and the importance of seed as a source of plant diseases have largely been neglected in southern Africa. Dry beans are the only crop in which a disease-free seed scheme and a routine seed-testing programme have been implemented in South Africa. In grain and legume crops used for human and animal consumption, emphasis of research is on toxin-producing, seedborne fungi. The role seeds play in introducing new diseases in developing countries is frequently overlooked. In this paper some of the recent and most important case studies will be discussed.

LEGUME AND GRAIN CROPS

Seedborne diseases of beans are well known in Africa. Some of the most important seedborne diseases in the highlands of eastern Africa are angular leaf spot (*Phaeoisariopsis griseola*), bean common mosaic virus, anthracnose (*Colletotrichum lindemuthianum*) and halo blight (*Pseudomonas syringae* pv. *phaseolicola*). Common blight caused by *Xanthomonas campestris* pv. *phaseoli* is important in the low-lying and tropical regions.

Anthracnose has always been regarded as the main seedborne fungal disease of beans in South Africa. However, in 1992, scab caused by *Elsinoë phaseoli* caused serious yield losses at two different localities (Phillips, 1994). Tracing these outbreaks to infested seed, scab has since been included in the list of diseases that must be absent during field inspection for the purpose of seed certification in South Africa. Yield losses of up to 70% were reported by Schwartz (1991) in Zambia. Scab is also regarded as the most destructive fungal disease of cowpeas in Zambia (Kannaiyan and Haciwa, 1993).

Halo blight and common blight are both associated with beans in South Africa. Bacterial brown spot, caused by *P. syringae* pv. *syringae*, however, took on epidemic proportions for the first time in the 1991/92 season (Serfontein, 1994). The disease has been spreading rapidly and is currently one of the most important diseases in commercial dry bean production in South Africa. The pathogen is regularly detected in certified seed lots. The importance of seed as a primary inoculum source has been confirmed by experimental comparison plantings with contaminated seed. Although bacterial brown spot is an important disease of large-scale commercial bean production, CIAT researchers have not encountered it to date in their bean programmes in the rest of Africa and South America (M.A. Pastor Corrales, personal communication).

Anthracnose on lupin, caused by *Colletotrichum gloeosporioides*, was detected for the first time in South Africa during 1993 in seed plantings of European origin. The disease was observed again on breeding lines in 1995. Plants were destroyed in both cases. Because *C. gloeosporioides* is generally known as a cosmopolitan pathogen with a wide host range, state restrictions on the organism are scant. However, colony morphology of the above-mentioned lupin strains differs considerably from that of other *C. gloeosporioides* strains isolated locally. Host origin and pathogenicity are currently being investigated (S.H. Koch, personal communication).

Although a fair amount of legumes and grains are produced in Africa, regular drought and increasing poverty necessitates the importation of vast quantities of both, but especially grains, by food aid schemes. These grains are mainly for human and animal consumption, not as seed for planting. Grain spillage during transportation and the use of the grain as seed can be a source of new seedborne diseases (Buruchara and David, 1995). Recently, shipments of wheat infected with *Tilletia indica*, the causal agent of Karnal bunt, were prevented from being imported into South Africa (E. Spies, personal communication). South African maize producers are also concerned about the possible introduction of *Pantoea stewartii* (*Erwinia stewartii*) in this way into the subcontinent.

RECURRENT AND NEW DISEASES ON OTHER CROPS

During the 1994/95 season, *Acidovorax avenae* subsp. *citrulli* was isolated for the first time from watermelon fruit with typical bacterial fruit blotch symptoms in South Africa. The watermelons were from an experimental trial planted with imported seed. Plants were immediately destroyed by removing and burning. No fruit blotch was observed during the subsequent season, suggesting that the swift action may have prevented further spread of the disease (J.J. Serfontein, unpublished data).

Prior to 1981, downy mildew of sunflower, caused by *Plasmopara helianthi*, was reportedly absent in South Africa (Sackston, 1981) and was also not detected in a survey of seedborne diseases in Mozambique done in 1987 (Tarp *et al.*, 1987). However, in 1991, downy mildew was reported as a new disease of sunflower in Zambia (Raemaekers *et al.*, 1991). In the 1992/93 and 1993/94 seasons the disease was widespread in seed production and commercial sunflower fields in Zimbabwe. During the 1993/94 season, the disease was also detected in sunflower fields from different seed sources in different localities in South Africa (Nowell, 1995). Seed produced in Zimbabwe during the 1992/93 season was destined for commercial sunflower production in South Africa. However, the occurrence of downy mildew resulted in a ban on importing of the seed into South Africa, with resultant major economic losses. Downy mildew is currently regarded as the most important new seedborne disease in Zimbabwe (C.M. Mguni, personal communication). Currently all sunflower seed imported to and exported from South Africa must be treated with metalaxyl (Van Wyk and Holtzhausen, 1994).

A new disease was recently found on crucifers in seedling nurseries in the Kwazulu-Natal province of South Africa. The disease was named 'chocolate spot' and a yellow pigmented bacterium was isolated from the seedlings (Tennant and Laing, 1996). Fatty acid profiles and sequencing showed the bacterium to be a *Xanthomonas* sp., but distinct from any of the four *X. campestris* pathovars attacking crucifers or any other known *X. campestris* pathovars. Further studies revealed that all the isolates of the organism concerned were parasitized by a temperate phage (Petersen *et al.*, 1996). The organism was isolated locally and by researchers overseas from crucifer seed produced in South Africa (M.D. Laing, personal communication). At this stage, the disease is still confined to nurseries in the eastern, more humid areas of South Africa.

Bacterial canker is a major disease of stone fruit in South Africa. Seeds of stone fruit trees are used for the production of rootstocks. After complaints about the poor germination of some peach seed batches, an investigation was launched (Roos *et al.*, 1989). *P. syringae* pv. *syringae* was isolated from a very high percentage of the seeds tested. This has important implications for the stone fruit industry, since seed was not previously recognized as a carrier of bacterial canker.

DISCUSSION

The introduction of new diseases into the subcontinent via seed is a major concern. State quarantine regulations of many countries apply only to new pathogens and not to new pathotypes or races of pathogens. A further problem is that once a pathogen is reported in the country, the state no longer regulates the importation of seed infected with the particular pathogen. This could lead to epidemics of diseases previously eradicated or diseases that are normally under control.

Africa is a net importer of grain. Although not imported as seed for planting, some of this grain is used as seed. Grain imported as food is often infected with seedborne pathogens. We have, for instance, isolated bacterial blight pathogens from dry beans imported from China. However, the importance of seed as a primary inoculum for diseases must be seen in perspective and McGee (1995) warned against the indiscriminate cataloguing of microorganisms on seed. Crop residues might well be the most important source of inoculum in crops like maize and wheat, which are mainly grown in monoculture. Most of the seed used for commercial agriculture is treated with fungicides, but in countries like Kenya cereal diseases like head smut of maize and loose smut of wheat remain a threat because some farmers plant seed from the previous crop and do not use chemically treated, certified seed. Recent information confirms that head smut of maize is still increasing in the cooler regions of Kenya (J.O. Onyango, personal communication).

Environmental conditions, cultural practices and infection levels of the seed may all influence the role of seed as a primary inoculum source. The seedbed and especially the seedling nursery phase used in the production of vegetables amplifies the effect of seedborne inoculum. 'Chocolate spot' of crucifers does not seem to have a noticeable influence on the yield; cosmetically, however, it is very important for the seed and seedling industries. In the subcontinent, where drier conditions tend to be the norm rather than the exception, the importance of some seedborne diseases is often overlooked. In normal or wetter years, sporadic epidemics of diseases like sclerotinia can often be traced to seed as the primary source. A different range of diseases affects small-scale farming and commercial farming, the latter having access to disease-free and chemically treated seed. Although breeding for resistance receives priority in most other African countries, emphasis in South Africa falls on the production of disease-free seed. In the case of beans, for instance, demand for certified seed is low in the major bean-producing countries of eastern and central Africa as a result of the high seed cost, unavailability of seed and because available certified seed has not necessarily been found to be disease-free (Buruchara and David, 1995).

The South African seed and agrochemical industries recently undertook a joint initiative concerning chemical seed treatment. One of their primary objectives is a national survey of seedborne diseases. Although similar surveys are currently being done in other southern African countries, information is not always freely available. The ISTA-PDC can address these problems by playing a key role in: (i) establishing a central, easily accessible and regularly updated database on seedborne diseases in the subcontinent; (ii) prioritising seedborne diseases that are of local importance and identifying research necessary on a regional basis, rather than for individual countries; and (iii) setting regulatory standards on the transportation of seed and grain across international borders.

REFERENCES

Buruchara, R and David, S. (1995) Seed quality: issues in small scale farmer bean production. In: *Proceedings: Fourth SADC Regional Bean Workshop, Potchefstroom, South Africa, 2–4 October 1995*. CIAT African Workshop Series no. 31, Appendix.

Kannaiyan, J. and Haciwa, H.C. (1993) Diseases of food legume crops and the scope for their management in Zambia. *FAO Plant Protection Bulletin* 41, 73–90.

McGee, D.C. (1995) Epidemiological approach to disease management through seed technology. *Annual Review of Phytopathology* 33, 445–466.

Nowell, D.C. (1995) The recent occurrence of sunflower downy mildew (*Plasmopara halstedii*) in Zimbabwe and South Africa. *South African Journal of Science* 91, xx (Abstract).

Petersen, Y., Freeborough, M.-J., Purves, M. and Qhobela, M. (1996) Chocolate spot disease of cabbage a possible bacterial–viral interaction. In: *Proceedings of the 34th Congress of the Southern African Society for Plant Pathology. University of Stellenbosch, 14–17 January 1996*, pp. 49–50 (Abstract).

Phillips, A.J.L. (1994) Occurrence of scab of *Phaseolus vulgaris* caused by *Elsinoë phaseoli* in South Africa. *Plant Pathology* 43, 417–419.

Raemaekers, R.H., Nawa, I.N., Chipili, J. and Sakala, A. (1991) Revised checklist of plant diseases in Zambia. General Administration for Development Cooperation, Brussels, Belgium. *Agricultural Editions* 18, 119 pp.

Roos, I.M.M., Hattingh, M.J. and Marasas, C.N. (1989) Transmission of *Pseudomonas syringae* pv. *syringae* by seed of stone fruit trees. In: Klement, Z. (ed.), *Proceedings of the 7th International Conference on Plant Pathogenic Bacteria. Budapest, Hungary, June 11–16, 1989*. Akademiai Kiado, Budapest, pp. 161–163.

Sackston, W.E. (1981) The sunflower crop and disease: progress, problems and prospects. *Plant Disease* 65, 643–648.

Schwartz, H.F. (1991) Scab. In: Hall, R. (ed.), *Compendium of Bean Diseases*. The American Phytopathological Society, St Paul, Minnesota, p. 25.

Serfontein, J.J. (1994) Occurrence of bacterial brown spot of dry beans in the Transvaal province of South Africa. *Plant Pathology* 43, 597–599.

Tarp, G., Lange, L. and Kongsdal, O. (1987) Seed-borne pathogens of major crops in Mozambique. *Seed Science and Technology*, 15, 793–801.

Tennant, G.W. and Laing, M.D. (1996) Studies on chocolate spot of crucifers in Kwazulu-Natal. In: *Proceedings of the 34th Congress of the Southern African Society for Plant Pathology. University of Stellenbosch, 14-17 January 1996*, p. 87 (Abstract).

Van Wyk, P.S. and Holtzhausen, M.A. (1994) Nuwe siekte van sonneblom nie noodwendig groot probleem nie. *Oliesadenuus* July 1994, 6–9.

Seedborne Disease in North America: Diversity and Change

J.W. Sheppard

Seed Borne Disease Unit, Central Seed Laboratory, Laboratory Services Division, Agriculture and Agri-Food Canada, Ottawa, Ontario, Canada K1A 0C6

INTRODUCTION

The diversity of climate and geography in North America present plant pathologists with a broad spectrum of plant disease problems. In many of our traditional crops, these diseases have been controlled or managed through cultural or chemical control programmes. Occasionally these programmes break down or are changed, allowing aggressive pathogens to re-emerge and thus causing excessive crop losses.

In recent years we have seen a shift away from these traditional crops as their production becomes less economically viable or, as in the case of tobacco (once a major cash crop second only to wheat in Canada), they fall from public favour. New crops and cropping practices move forward to fill this void, bringing with them new disease pressures.

The world today is shrinking rapidly. Seed production is no longer limited to the same region where the final crop is produced. International movement of seed, whether in commerce or as part of seed multiplication programmes, is commonplace. This international movement of seed brings increased risk of disease introduction to cropping areas.

The purpose of this review is not to provide an inventory of seedborne diseases of North America, but rather to provide an overview of their importance and to examine the role that the ISTA-PDC should be playing in addressing these problems.

 Seed Health Testing (eds J.D. Hutchins and J.C. Reeves)

RECURRENCE OF PREVIOUSLY CONTROLLED DISEASES

Barley Stripe

Barley stripe is an enigma for plant pathologists in Canada. It should be widespread yet, with the exception of the Peace River Valley of Northern Alberta, occurs very infrequently. It was first recorded in Canada in 1913. Severe outbreaks of this disease were reported in 1920, 1930, 1940, 1953 and 1954. In 1977 it reappeared after an absence of 20 years. It was reported only twice, once in the Maritimes and once on the Prairies. The apparent absence of this disease in Canada between 1959 and 1977 has largely been attributed to the use of mercurial seed dressings. Surveys carried out in 1959 and 1961 indicated that 75% of seed sown was treated. By 1976 only 8.2% of the seed sown was treated. This change in seed treatment applications has been attributed to three factors: (i) increased breeding for resistance to true loose smut; (ii) restrictions on the use of mercurial compounds for seed treatments since 1970; and (iii) changes in treatment applications from blanket treatments to selective treatments based on cultivar susceptibility and increase of disease. In spite of the reduction in use of seed treatments and a grower preference for later maturing cultivars which require early seeding to ensure maturity, the disease appears in only a few fields each year and causes no significant losses outside the Peace River Valley (Tekauz and Chilko, 1979). In the Peace River valley, however, annual surveys show 25% of the fields to be infected and incidence within fields to be as high as 20% (A. Tekauz, personal communication).

Anthracnose on Bean

Colletotrichum lindemuthianum (Sacc & Magn.) Briosi & Cav. is an important disease of field bean (*Phaseolus vulgaris* L.). In 1976 bean anthracnose caused by race δ and, to a lesser degree, by race λ became epidemic in Ontario (Wallen, 1976, 1979; Tu and Aylesworth, 1979; Tu, 1988). In 1977 a breeding programme was initiated to transpose a resistance gene (*Are*) to recommended cultivars. In addition a strict programme of seed treatment with benzimidazoles and field inspection of all pedigreed seed plots for bean anthracnose was initiated in Ontario. The disease, caused primarily by the α and δ races, was last observed in commercial fields during the 1983 growing season (Tu *et al.*, 1984). Surveys during the summer of 1993 found anthracnose in six of nine locations in south-western Ontario (Tu, 1994a). Various bean lines were affected including those with the *Are* gene which should have been resistant to the α, β, γ, δ, λ and ε races (Tu *et al.*, 1984). Subsequent investigations revealed that the 1983 anthracnose outbreak was caused by a new race, α-Brazil, introduced into Canada from Michigan (Tu, 1994b). In 1995 the ε race was again observed in several pedigreed seed fields in Ontario (J. Sheppard, unpublished) and in Michigan (Balardin and Kelly, 1996).

Common Bacterial Blight of Bean

Disease management began as a result of an epidemic of bacterial blight (*Xanthomonas campestris* pv. *phaseoli* (E. F. Smith) Dye.) in 1966 which resulted in economic loss and a reduction of seed quality. The principal effort for management of this disease was the select seed programme for field beans. This programme supplied growers with seed from Idaho which is either blight-free or has a very low level of bacterial blight. This breeder seed is distributed to approximately 50 select seed growers who produce up to 1 hectare each for the production of select seed. Seed from these plots is grown the next year, usually on the same farm, to produce foundation seed. The final pedigreed seed produced is certified and from this class commercial seed is produced.

Initially the programme was based on a zero tolerance for bacterial blight based on a field inspection and laboratory test in select plots. A trace amount of blight was permitted in foundation seed. Select plots were inspected twice during the growing season. Foundation and certified fields were inspected once. The programme operated in this manner for over 15 years and was effective in reducing the incidence of blight and subsequent losses to acceptable levels (Wallen and Galloway, 1979). During the early 1980s a series of government cutbacks and downsizing resulted in a weakening of the Canadian select seed programme. Today, seed from Idaho is still used to produce select plots. Plot inspection, however, has been reduced to a single inspection directed towards genetic purity rather than disease detection. Disease is only noted as a comment and inspection may not take place at optimum times for disease observation. Laboratory testing is performed on request from the grower and for a fee. The result has been a gradual increase in the incidence of blight in select plots and subsequent crops. In 1980 and again in 1981, 205 of the select plots were rejected because of bacterial blight (Sheppard, 1983). By 1995, 19 out of 20 select plots examined in the London, Ontario, region were found to have significant levels of bacterial blight. Blight was noted during field inspection in 67 of 68 foundation and certified seed fields.

No information is available to relate actual crop losses with this apparent increase in the incidence of bacterial blight in Ontario seed fields. A recent survey of seed growers conducted by the Canadian Seed Growers' Association indicated that, from a grower's perspective, white mould, caused by *Sclerotinia sclerotiorum*, was more significant than bacterial blight. This increase in bacterial blight and other bean diseases is not limited to Ontario: in neighbouring Michigan, these diseases have been increasing at an alarming rate. In an effort to manage and reduce the incidence of disease in field bean, Michigan has made changes to its seed law. These changes require that field bean seed produced east of a line dividing the Central and Mountain time zones and sold or offered for sale in Michigan be field inspected and laboratory tested for bacterial blights, anthracnose and other diseases deemed to be a threat to the bean industry (Anon., 1996).

Tan Spot

Tan spot caused by *Pyrenophora tritici-repentis* (Died.) Drechs. (syn *P. trichostoma* (Fr.) Wint.) anamorph *Dreschslera tritici-repentis* (Died.) Shoem. occurs worldwide and has in recent years become an economically important disease of wheat. The increased incidence is largely attributed to the shift towards stubble retention generally associated with soil conservation practices. Tan spot and the septoria complex caused by *Septoria nodorum* and *S. tritici* are considered to be two most important leaf spot diseases in the Canadian Prairies (A. Tekauz, personal communication). Incidence of tan spot is expected to increase as conservation tillage practices are more widely adopted by farmers.

EMERGENCE OF NEW AND POTENTIALLY YIELD-LIMITING DISEASES

Karnal Bunt

Karnal bunt (*Tilletia indica*), also known as new bunt or partial bunt, was first reported on wheat from the Karnal region of northern India (Mitra, 1931). Presence of the disease in the Sonora region of Mexico was confirmed in 1972 (Duráin, 1972; Duráin and Cromarty, 1977). By 1984 the disease had become well established in Mexico and the Mexican government imposed regulations to control the spread of the disease (Delgado, 1984; Lira, 1984).

On 8 March 1996, scientists with USDA's Agricultural Research Service (ARS) confirmed the presence of Karnal bunt in Arizona in certified durum wheat seed, the first occurrence of this disease in the USA. Suspect seed samples had been detected at a seed dealership a few days earlier during routine testing by the Arizona Department of Agriculture. Emergency action notices were instituted on infected properties, seed, farm equipment, planted wheat, and soil associated with the infected wheat. State and Federal quarantines were put into place to augment this emergency action, and USDA established a wheat export certification team to develop options for dealing with potential trade issues.

Although the overall crop losses caused by Karnal bunt might not be severe, the disease has quarantine significance and thus could affect US grain exports. The US exports one-third of world wheat exports with US exports in the calendar year 1995 valued at $5.5 billion.

Watermelon Blotch

In the spring of 1989 a new bacterial disease of watermelon was observed in production fields in the USA (Hopkins, 1989). The fruit blotch bacterium, *Acidovorax avena* subsp. *citrulli* (Schaad *et al.*) Willems *et al.*, causes a seedling blight, leaf lesions and fruit symptoms. In the spring of 1994 severe outbreaks of the disease in transplant production facilities in the southern USA led to litigation between growers, who suffered severe losses, and the seed companies. The risk of lawsuits and pending litigation resulted in a suspension of watermelon seed in the USA which threatened the existence of the 1995 crop.

The seed companies, transplant growers, watermelon producers, private and public researchers have pooled their resources to conduct research and to explore disease management options. Today all watermelon seed is tested using a 10,000 seed grow-out test under conditions favourable for disease development. Watermelon seed now carries a statement on the label to the effect that the seed has been tested and found negative for fruit blotch pathogen, but the absence of the pathogen cannot be guaranteed (Latin and Hopkins, 1995).

Bacterial fruit blotch of watermelon appears to be under control, but not without lingering costs to the industry and ultimately the consumer. Seed testing requirements and changes to seed handling and packaging have increased seed costs significantly. Global concern over this disease has resulted in some countries requiring vegetable seed of several non-cucurbit species such as eggplant and cabbage to be tested and found free from *Acidovorax*.

Specialty Crops

Interest in production of specialty crops has been increasing. Ginseng for instance has been grown in Canada for over 100 years. A perennial herb, the plant is normally grown for 4 years before harvest. Dried roots are exported to Asia where they are used for medicinal purposes. Recently there has been an expansion in acreage and the farm gate value of this crop has increased from $3 million in 1982 to $50 million in 1993. Losses due to disease are high with plants affected at all stages of growth. Infection at the seedling stage due to seed or soil-borne disease can result in irregular plant development which ultimately may affect the root shape and size in surviving plants. In Canada there has been increased interest in the production of herbs such as basil in field and in greenhouses. *Fusarium* wilt, probably introduced to North America with seed from Italy in 1989, has become a major disease problem for this emerging industry.

CONCLUSION

These are but a few examples of current seedborne disease problems in North America. As our governments and industries accept the economic realities of the new millennium it is obvious that we can no longer rely on any one sector of the economy to bear the costs of disease control or management. A cooperative effort of all involved is required. Seed health standards remain an important tool in disease management but must be realistic and based on sound scientific data backed up with efficient and reproducible test methods. The Plant Disease Committee of ISTA must play a lead role in guiding activities of programmes such as the ISHI through development, validation and publication of standardized test methodologies and encourage the use of these methods in international seed trade.

REFERENCES

Anonymous (1996) Enrolled Senate Bill No. 713. State of Michigan 88th Legislature Regular Session of 1996 Section 7a.

Balardin, R.S. and Kelly, J.D. (1996) Identification of race 65-epsilon of bean anthracnose (*Colletotrichum lindemuthanium*) in Michigan. *Plant Disease* 80, 712

Delgado, S. (1984) Mexican phytosanitary policy in relation to Karnal bunt. In: *Karnal Bunt Disease of Wheat. Proceedings of a Conference, April 16–18 1984, Ciudad Obergon, Sonora, Mexico.* CIMMYT, p. 20.

Duráin, R. (1972) Aspects of teliospore germination in North America smut fungi II. *Canadian Journal of Botany* 50, 2569–2573.

Duráin, R. and Cromarty, R. (1977) *Tilletia indica*; a heterothallic wheat bunt fungus with multiple alleles controlling compatibility. *Phytopathology* 67, 812–815.

Hopkins, D.L. (1989) Bacterial fruit blotch of watermelon: a new disease in the Eastern USA. *Proceedings of Cucurbitaceae* 89, 74–75.

Latin, R.X., and Hopkins, D.L. (1995) Bacterial fruit blotch of watermelon: the hypothetical exam question becomes reality. *Plant Disease* 79, 761–765.

Lira, M. (1984) The Karnal bunt situation in northwest Mexico. In: *Karnal Bunt Disease of Wheat. Proceedings of a Conference, April 16–18 1984, Ciudad Obergon, Sonora, Mexico.* CIMMYT, pp. 24–26.

Mitra, M. (1931) A new bunt on wheat in India. *Annals of Applied Biology* 8, 178–179.

Sheppard, J.W. (1983) Historical perspectives of the production of disease-free seed, control and management of bacterial blights of beans in Canada. *Seed Science and Technology* 11, 885–891.

Tekauz, A. and Chilko, A.W. (1979) Leaf stripe of barley caused by *Pyrenophora graminea*. Occurrence in Canada and comparison with barley stripe mosaic. *Canadian Journal of Plant Pathology* 2, 152–158.

Tu, J.C. (1988) Control of bean anthracnose caused by delta and lambda races of *Colletotrichum lindemuthianum*. *Plant Disease* 72, 5–8.

Tu, J.C. (1994a) Reoccurrence of anthracnose disease on field bean in southwestern Ontario in 1993. *Canadian Plant Disease Survey* 74, 96.

Tu, J.C. (1994b) Occurrence and characterization of the alpha-Brazil race of bean anthracnose (*Colletotrichum lindemuthanium*) in Ontario. *Canadian Journal of Plant Pathology* 16, 129–131.

Tu, J.C. and Aylesworth, J.W. (1979) An effective method of screening white (pea) bean seedlings *Phaseolus vulgaris* L. for resistance to *Colletotrichum lindemuthanium*. *Phytopathologische Zeitschrift* 99, 131–137.

Tu, J.C., Sheppard, J.W. and Laidlaw, D.M. (1984) Occurrence and characterization of the epsilon race of bean anthracnose in Ontario. *Plant Disease* 68, 69–70.

Wallen, V.R. (1976) Anthracnose on field beans in Ontario. *Canadian Plant Disease Survey* 56, 109.

Wallen, V.R. (1979) The occurrence of the lambda race of bean anthracnose in Ontario. *Canadian Plant Disease Survey* 59, 69.

Wallen, V.R. and Galloway, D.A. (1979) Effective management of bacterial blight of field beans in Ontario – a 10 year program. *Canadian Journal of Plant Pathology* 1, 42–46.

New and Priority Seedborne Diseases in South America

8

M.F. Batista

EMBRAPA/CENARGEN, SAIN Parque Rural, PO Box 02372, CEP 70.970-900, Brasilia DF, Brazil

INTRODUCTION

In 1989 the Comite de Sanidad Vegetal del Cono Sur (COSAVE) considered establishing sanitary and phytosanitary agreements to be signed by Argentina, Brazil, Chile, Paraguay and Uruguay. In the last 6 years these five countries have worked together to try to establish standard criteria to be adopted for commercial trading. The agreement also considered the need to respect international notification procedures as approved in World Trade Organization Phytosanitary and Sanitary Measures agreements.

COSAVE has prepared a list of quarantine pathogens (Diário Oficial, 1996) which includes both A1 and A2 pathogens. A1 pathogens are those which have never entered the country while A2 pathogens are those which have already entered the country but are under official control and restricted to limited areas. The A1 and A2 lists for this region are shown in Tables 8.1 and 8.2 respectively. In 1993, considering the same needs, the Cartagena Agreement was signed by Bolivia, Colombia, Equator, Peru and Venezuela. The Andean Group Countries (GRAN) similarly prepared a list of pests (Acuerdo de Cartagena, 1993) which are exotic to the five countries (Table 8.3).

Introduction of plants has been one of the most effective actions for agricultural development in the tropics. Examples of successful introductions include coffee and soybean in Brazil, banana in South and Central America, sugarcane in South America and the Caribbean, and grass pastures in Latin America. The development of improved varieties requires germplasm exchange between research institutions throughout the world. A considerable

Table 8.1. A1 list of seedborne quarantine pathogens for the COSAVE region.

Pathogen	Hosts
Phoma tracheiphila	*Citrus* spp.
Clavibacter michiganensis ssp. *insidiosus*	alfalfa, clover
Spiroplasma citri (Stubborn)	*Citrus* spp.
Xanthomonas campestris pv. *oryzae*	rice
Xanthomonas campestris pv. *oryzicola*	rice
Xanthomonas campestris pv. *citri* (biotypes D and E)	*Citrus* spp.
Citrus tristeza virus (severe races)	*Citrus* spp.
Potato spindle tuber viroid	tomato

Table 8.2. A2 list of seedborne quarantine pathogens for the COSAVE region.

Pathogen	Country	Host
Xanthomonas campestris pv. *citri* (biotype A)	Argentina, Brazil, Paraguay, Uruguay	*Citrus* spp.
Xanthomonas campestris pv. *citri* (biotype B)	Argentina, Paraguay	*Citrus* spp.
Xanthomonas campestris pv. *citri* (biotype C)	Brazil, Paraguay	*Citrus* spp.
Pseudomonas solanacearum (race 3)	Chile, Argentina, Brazil, Paraguay, Uruguay	potato, tomato, pepper
P. syringae pv. *phaseolicola*	Argentina	bean
Erwinia stewartii	Paraguay	maize
Potato spindle tuber viroid	Argentina	potato
Tilletia indica (*Neovossia indica*)	Brazil	wheat, triticale, *Agropirum* spp., *Festuca* spp.
Tilletia controversa	Argentina, Uruguay	wheat, barley
Tilletia barclayana	Brazil	rice
Phialophora gregata	Brazil	soybean

Table 8.3. List of exotic seedborne pathogens for the Andean subregion.

Pathogens	Host plants
Xanthomonas campestris pv. *citri*	*Citrus* spp.
Corynebacterium michiganense pv. *sepedonicum*	potato, tomato, eggplant
Apple mosaic virus	apple, almond
Sphacelotheca erianthi	sugarcane
Glomerella cingulata	coffee, mango, apple, tomato, *Annona*, cocoa
Urocystis agropiri	wheat
Sphaceloteca cruenta	sorghum
Sclerophthora macrospora	rice, oat, barley, maize, wheat, rye

and ever-increasing interest is being shown by virtually every country for the establishment of germplasm banks. This is considered essential to preserve primitive genetic resources for future use and has been encouraged by the International Board for Plant Genetic Resources (IBPGR). The movement of germplasm involves the risks of introduction of pests, pathogens and weeds which are sometimes carried by plants or plant materials. There are many examples of disastrous diseases introduced into several countries. Brazil is not an exception: citrus canker caused by *Xanthomonas campestris* pv. *citri* was introduced in 1957 and cost over $5 million dollars to eradicate, but is still present in São Paulo and in other parts of the country (Galli *et al.*, 1968).

In the last two decades agricultural research in Brazil has improved substantially as a result of pressures for new and modern technology to increase production and productivity. In this context the improvement of genetic resources has arisen as an important component of the modernization of Brazilian agriculture. As a consequence of the increase in the introduction of genetic material into Brazil there has been a significant increase of the risk of exotic pests and pathogens entering the country (Rocha, 1984). The government of Brazil had taken legislative steps, as far back as 1934, in order to prevent the introduction of exotic pests and diseases and to limit their spread (Alves, 1977, 1979). Modifications in the legislation have been introduced as conditions changed and new facts became available, resulting in further restrictions or suppression of requirements found not to be necessary.

Brazil is a signatory to the International Plant Protection Convention established by the FAO in 1951 (Anon., 1976). Under these regulations, rules governing the import and export of plants and plant products, insects, mites and microorganisms are established. Examples of similar agreements include the Cartagena Agreement, and the Phytosanitary requirements for the GRAN countries. COSAVE have established criteria for the harmonization of procedures and methods for phytosanitary diagnosis and for phytosanitary products. Both COSAVE and GRAN are trying to follow the same steps of NAPPO and EPPO and one of their main targets is to minimize the use of phytosanitary measures as barriers to international trade, as well as reducing the risk of introduction of new diseases into a country. Selected examples of pathogens of economic importance are discussed below.

ECONOMIC IMPORTANCE OF THE A1 QUARANTINE SEEDBORNE PATHOGENS

Phoma tracheiphila (Petri) Kanchaveli & Gikashvili

Phoma tracheiphila causes a vascular disease known as *mal secco* (from the Italian *male*, disease, and *secco*, withered, dry). *Mal secco* is the most destructive fungal disease of lemon and other susceptible *Citrus* species in the

Mediterranean region. Losses arise from a heavy reduction in yield due to withering of infected twigs and branches and the severe pruning that growers carry out in order to control the disease. Moreover, whole trees may be killed by the disease. If they are replaced by younger healthy trees, the orchard becomes uneven in age and size of trees. It is hard to estimate the average losses due to *mal secco*, because the disease incidence changes from year to year even in the same orchard (Smith *et al.*, 1988).

Clavibacter michiganensis Subsp. *insidiosus* Davis *et al.*

The disease this pathogen causes is economically important only in lucerne (*Medicago sativa*), although *Medicago falcata*, *Melilotus alba* and *Onobrychis viciifolia* are also said to be susceptible. Symptoms are accompanied by chlorosis, reduction in size and often cupping of leaflets. Marginal, papery, whitish grey necrosis of leaflets is common and wilting may occur during hot weather. However, in North European conditions the characteristic symptoms are not those of a wilt but of stunting and proliferation of stems (Smith *et al.*, 1988).

Spiroplasma citri Saglio *et al.*

Affected citrus trees appear slightly to severely stunted, frequently with abnormally dense bunches and upright foliation. Excessive dieback of shoots and branches occurs in severe cases. Development of multiple buds results in witches' broom; affected trees have low yields. Stubborn disease is destructive in most countries that grow citrus under hot, dry, desert or semi-desert conditions. Fruit quality is inferior and many fruits are malformed. Production from a diseased tree can be reduced by 50–100% (Smith *et al.*, 1988).

Xanthomonas campestris pv. *citri* (Hasse) Dye

This pathogen causes citrus canker in the Far East and parts of Africa and South America, and has been introduced several times into Florida, USA where it is currently again under eradication. It is a major quarantine organism for Mediterranean citrus (Smith *et al.*, 1988).

Citrus tristeza virus

Citrus tristeza virus (CTV) has caused extensive damage to citrus crops in Spain and threatens citrus production of sour orange rootstocks in Israel and other Mediterranean countries. Worldwide during the last 50 years it has been estimated that about 40 million trees were lost or became unproductive due to CTV (Smith *et al.*, 1988).

ECONOMIC IMPORTANCE OF THE A2 QUARANTINE SEEDBORNE PATHOGENS

Pseudomonas solanacearum (Smith) Smith

The disease limits the growth of many important tropical and subtropical host crops but in Europe it has caused substantial loss only in potato, and then only in Portugal, Greece and Yugoslavia (Smith *et al.*, 1988) (see Chapter 4).

Pseudomonas syringae pv. *phaseolicola* (Burkholder) Young *et al.*

In North America, where most cultivars carry some resistance to halo blight, the disease, although less important than common blight, is still a major disease of beans, causing an estimated loss in 1976 of $2 million. In Europe, where cultivars are more susceptible, it is probably the most important disease of dwarf and runner beans. Yield losses range from 12.5% to 43% (Smith *et al.*, 1988).

Erwinia stewartii (Smith) Dye

E. stewartii causes Stewart's disease or bacterial wilt of maize, a serious North American disease that can cause total crop loss of sweetcorn. It also affects dent, flint, flour and popcorn cultivars to a lesser extent. Damage has been more extensive in Italy, where the entire plant is attacked, than in the USA where the disease has been established and economically serious for many years, and only the leaves are attacked in mature plants. Although the bacterium may be transmitted in seed and occasionally overwinters in soil, manure or plant debris, this is of little importance compared with transmission by insect vectors in which the bacteria survive during hibernation and are then transferred from plant to plant in the growing season (Smith *et al.*, 1988).

Tilletia indica Mitra (syn. *Neovossia indica* (Mitra) Mundkur)

This causes Karnal bunt of wheat. It is principally seedborne and could cause serious problems in most warm temperate parts of the world (Europe, North America, Australia) if it spread. It is universally considered as a serious quarantine organism. Its accidental introduction to Mexico has highlighted the dangers from the exchange of germplasm between research and breeding stations in different continents (Smith *et al.*, 1988).

Tilletia controversa Kühn

T. controversa causes dwarf bunt of winter wheat and some grasses. The disease is currently important mainly in Eastern Europe (Hungary, Poland, former USSR), and also in North America. Although the fungus can be seedborne, infection typically arises from spores in the soil, which can remain viable for up to 10 years (Smith *et al.*, 1988).

ECONOMIC IMPORTANCE OF THE EXOTIC SEEDBORNE PESTS FOR THE ANDEAN SUBREGION

Corynebacterium michiganense pv. *sepedonicum* Dye & Kemp

For many decades ring rot was the major potato disease in the USA and Canada with estimates of annual losses as high as 10–15% of total crops a year in some US states. Strenuous and costly efforts in recent years have reduced its importance there although it is still considered a major disease (Smith *et al.*, 1988).

Glomerella cingulata (Stoneman) Spaulding & v. Schrenk

Due to its required temperature range for growth, *G. cingulata* will not rot fruit in low-temperature stores and so has little practical importance in Europe. It is of greater importance in the warmer wetter summers of North America, and specially in subtropical conditions (in Brazil it attacks fruit even in the orchard). Economic importance differs from one country to another and may also vary from one year to another. In Portugal losses higher than $3 million were registered in 1965. More recently serious losses were observed too; for example, in areas of high relative humidity more than 50% of fruit production may be affected (Smith *et al.*, 1988).

REFERENCES

Acuerdo de Cartagena (1993) *Propuesta de Norma y Manual de Requisitos y Procedimientos de Cuarentena Vegetal para el Comercio Agricola Subregional y con Terceros Paises.* 362 pp.

Alves, H.T. (1977) Sistema Brasileiro de defesa sanitária vegetal. *Fitopatologia Brasileira* 2, I–V Noticiário.

Alves, H.T. (1979) *Legislação Fitossanitária Brasileira.* Ministério da Agricultura, Secretaria Nacional de Defesa Agropecuária, Brasília.

Anonymous (1976) *Regulamento de Defesa Sanitária Vegetal 3.* Ministério da Agricultura, Departamento Nacional de Produção Vegetal, Divisão de defesa Sanitária Vegetal. Brasília.

Diário Oficial (1996) *Suplemento AO* no. 58, 25 March 1996, National Press, 72,pp.

Galli, F., Tokeshi, H., Carvalho, P., Balmer, C.T., Kimati, H., Cardoso, C.O.N. and Salgado, C.L. (1968) *Manual de Fitopatologia: doenças das plantas e seu controle.* Agronômica Ceres, São Paulo.

Rocha, H.M. (1984) *Plant Introduction and Quarantine in Brazil. ARNPC/IRRDB Workshop on SALB.* CEPLAC, Itabuna, Bahia, Brazil.

Smith, I.M., Dunez, J., Lelliott, R.A., Phillips, D.H. and Archer, S.A. (1988) *European Handbook of Plant Diseases.* Blackwell Scientific Publications, London, 583 pp.

Seedborne Diseases in Australia and New Zealand

9

M. Ramsey[1], E. Alberts[1], J. Fletcher[2] and N. Grbavac[3]

[1]South Australian Research and Development Institute (SARDI), GPO Box 397, Adelaide, South Australia, Australia, 5001; [2]Crop and Food Research, Private Bag 4704, Christchurch, New Zealand; [3]Ministry of Agriculture and Fisheries, PO Box 609, Palmerston North, New Zealand

INTRODUCTION

Seedborne pathogens affect both the agricultural and horticultural industries in Australia: (i) by introducing and spreading exotic diseases; (ii) by providing inoculum sources for endemic disease outbreaks; and (iii) in marketing seed, both nationally and internationally.

Quarantine regulations are enforced to prevent spread of seedborne diseases. Australian quarantine laws restrict the importation of some seed to reduce the risk of introducing specific exotic pathogens. Within Australia, state laws further affect the movement of seed. Quarantine regulations imposed by other countries affect export markets for seed.

Endemic seedborne diseases are of significant concern to pathologists, government and farming industries but there are no current statutory requirements for pathogen freedom under seed certification legislation. However, plant breeders and seed merchants have a 'duty of care' to ensure that any seed is suitable for sale under Australian common law (Bertholini, 1995). In response, voluntary testing is conducted by an increasing number of plant breeders and seed merchants to meet market demand for pathogen-tested seed.

INTRODUCING AND SPREADING DISEASES IN AUSTRALIA

Australia's agricultural industries benefit by being free from a number of major crop diseases. Stringent quarantine regulations are enforced on the

 Seed Health Testing (eds J.D. Hutchins and J.C. Reeves)

importation of seed to prevent the introduction and spread of exotic diseases into the country, although seed is generally a lower risk than other planting material (Chandrashekar and Culvenor, 1993). Regulations vary between crops, however, and are based on an assessment of the importance and likely impact of exotic seedborne diseases, and the value of the crop. In addition to national controls, states also have powers to impose controls on movement of seed.

An outbreak of lupin anthracnose (*Colletotrichum gloeosporioides*) on white lupins (*Lupinus albus*) in Western Australia during 1994 followed unrestricted importation of breeders' seed from Europe. Anthracnose is the major seedborne disease on white lupins in France, Chile and Brazil (Gondran *et al.*, 1994); however, all lupins are susceptible. Although the outbreak was eradicated, the disease occurred again in Western Australia, South Australia and New South Wales during 1996. This poses a serious threat to the large Australian lupin industry with one million tonnes produced annually (Lazenby *et al.*, 1994). An eradication strategy is being formulated and seed test development is an urgent priority.

Ergot (*Claviceps africana* or *C. sorghi*) in sorghum was recently identified in Queensland. Sorghum is a major cereal crop in tropical Australia with annual grain production of 1.2 million tonnes (Lazenby *et al.*, 1994). Sorghum is a restricted seed in Australia (Anon., 1992) and the source of the current outbreak is unknown. Ergot is disseminated as sclerotes contaminating seed (Ahmed and Reddy, 1993) and local regulations may be imposed to restrict further spread.

MAJOR ENDEMIC SEEDBORNE DISEASES IN AUSTRALIA

Although many diseases are seedborne, only those pathogens where seed constitutes the major source of infection in new crops are considered here.

Agricultural Crops

Pulses

Among the widely grown agricultural crops in Australia, the pulses have the most severe problems with seedborne pathogens. Of particular concern are the *Ascochyta* blight diseases which affect field peas (*Pisum sativum*), faba beans, lentils (*Lens culinaris*), chickpeas (*Cicer arietinum*) and vetches (*Vicia* spp).

Ascochyta lentis is the most important disease on lentil. The major sources of infection are infected seed and crop residues (Gossen and Morrall, 1981; Kaiser and Hannan, 1986); however, as lentils are a new crop in Australia, there are few infected residues and seed is the main source of infection. Spread from seed is most prevalent when crops are sown into cool, wet soils (Gossen and Morrall, 1981, 1986). Although lentil seed infection levels of 49% were recorded in some samples in 1994, conditions during 1995 did not favour disease development. However, this high level of seed infection coincided with the release of a new cultivar and highlighted the legal implications for breeding programmes and merchants selling infected seed. Based on this experience, voluntary testing for *A. lentis* is now widely practised.

The presence of *A. rabiei* in chickpeas in Australia was recently confirmed (Khan and Ramsey, unpublished data); however, its distribution is unknown. The chickpea industry is expanding rapidly into new areas and seed infection is an important risk (Haware *et al.*, 1986; Ketalaer *et al.*, 1988; Ahmed and Reddy, 1993). Western Australia has restricted chickpea seed imports from other states to protect their industry. During 1996, the Grains Research and Development Corporation (GRDC) supported a survey of seed. Fifty-seven seed lots were tested. No infection was detected in 56 seed lots harvested in 1995 when conditions were dry at harvest, whereas *A. rabiei* was identified in a single sample harvested under wet conditions experienced in 1992. Seed infection for several fungal species, including *A. fabae*, is correlated with the amount of rainfall during the season to establish the disease in the crop and then during maturation to allow infection of the seed (Diekmann, 1993). Based on the test results, seed infection posed a negligible threat to chickpea crops during 1996.

Other *Ascochyta* blight diseases, including *Mycosphaerella pinodes* in field pea and *A. fabae* on faba beans, are also highly seedborne. However, these diseases are now endemic and stubble retention, which minimizes soil erosion, and dry conditions over summer facilitate survival of the pathogens in the field. These stubbles are important alternative sources of infection. Seed tests are recommended for seed merchants and for growers in areas where the diseases are not well established.

Pea seedborne mosaic virus (PSbMV) has been detected in commercial pea seed and within breeding programmes (Alberts, unpublished data); however, its importance in crops has not been determined. PSbMV testing is now applied to pea breeding lines near release, using a modification of the dot immunobinding assay (DIBA) (Ligat *et al.*, 1991). PSbMV is seedborne in peas and lentils but also infects chickpeas and *Vicia* spp. PSbMV is a quarantinable disease in Australia and most pulse germplasm is rigorously screened, yet lentils are imported without restriction. This anomaly poses a constant threat to Australian pulse industries.

Cucumber mosaic virus (CMV) is the major seedborne disease in lupins (*L. angustifolius*) and seed infection levels above 0.1% are sufficient to cause epidemics in some seasons (Geering, 1992). Growers are advised to sow only tested seed (Alberts *et al.*, 1985) to minimize risks. The disease problem in lupins is sometimes compounded by mixed infections of CMV with alfalfa mosaic virus (AMV) (Nutter *et al.*, 1996). AMV is not seedborne in lupins but spreads rapidly from neighbouring medic (*Medicago* spp.) pastures and is lethal to lupins in mixed infection with CMV. During 1995, seedborne CMV was also detected in commercial lentil seed lines (E.V. Alberts, unpublished data) and routine seed testing has been adopted by some producers.

Bacterial blight (*Pseudomonas syringae* pv. *pisi*) caused significant crop losses in 1992 during an unusually wet season. A seed test using a modified ELISA system has been developed and is commercially available to industry (Gooden and Alberts, 1995).

Other agricultural crops

Bacterial blight (*P. syringae* pv. *coriandricola*) was recently identified in coriander (*Coriandrum sativum*) crops (Dennis *et al.*, 1995). Epidemics occurred in 1992 and 1993 and have devastated the industry. Bacterial blight reduces both the yield and quality of seed. The disease is also reported from Europe (Taylor and Dudley, 1980; Mavridis *et al.*, 1989; Toben and Rudolf, 1996) and North America (Cooksley *et al.*, 1991). Seed tests, using a modification of the method for *P. syringae* pv. *phaseolicola* in beans (van Vuurde and van den Bovenkamp, 1989), show that most coriander seed in Australia is infected. Current research focuses on identifying sources of pathogen-free seed.

The smut and bunt diseases (*Tilletia*, *Ustilago* and *Urocysistis* spp.) are major problems in the cereal industries in Australia, even though they generally occur at low levels in crops. The significance of smut diseases is amplified by the requirements for disease freedom in export markets. The diseases are routinely controlled by the application of seed treatments; however, this is an important added expense for growers. In future, testing to identify seed contaminated with smut spores may reduce dependence on fungicide treatments.

Horticultural Crops

Vegetatively propagated crops

Seedborne diseases in most horticultural crops in Australia are poorly studied. In general there are more problems with vegetatively propagated crops than there are for crops grown from true seed. Important examples include garlic yellow streak virus (GYSV) and garlic mosaic virus (GMV) which are common problems for the garlic industry; in potatoes, the removal of mercurial fungicides has resulted in an increased prevalence of a number of fungal diseases, including powdery scab (*Spongospora subterranea*), gangrene (*Phoma exigua*), rhizoctonia (*Rhizoctonia solani* AG3), silver scurf (*Helminthosporium solani*) and black dot (*Colletotrichum coccoides*). In addition, bacterial wilt (*Pseudomonas solanacearum* biovar 2) is a potential threat (T.J. Wicks, personal communication).

Diseases spread on true seeds

Brassica diseases including black rot (*Xanthomonas campestris*) and blackleg (*Leptosphaeria maculans*) are re-emerging as problems for the industry. These diseases are normally controlled by using a hot water (Persley *et al.*, 1989) or aerated steam treatment of seed. However, many growers are reluctant to use the recommended treatment due to inconvenience and the effect on seed viability (T.J. Wicks, personal communication).

Seed is an important source of infection for target spot (*Alternaria solani*), tomato mosaic virus (TMV), bacterial canker (*Clavibacter michiganensis* subsp. *michiganensis*), bacterial spot (*X. campestris* pv. *vesicatoria*) and bacterial speck (*P. syringae* pv. *tomato*) in tomato (Persley, 1994). In recent years, acid seed extraction techniques, which control TMV and surface-borne contamination, have been adequate (Persley *et al.*, 1989). However, these diseases remain a risk because infection can be internal in the seed, and control of this requires an additional hot water treatment (Persley *et al.*, 1989).

Watermelon blotch (*Acidovorax avenae* subsp. *citrulli*) has recently emerged as a problem in Queensland. This disease had rarely occurred for a number of years (Persley, 1994), but has re-emerged following a wet season (R.G. O'Brien, personal communication). Seed contamination is now a significant threat for the next season's crops.

Neck and bulb rot (*Botrytis* spp.) is important in onions although most seed is treated with benzimidazole fungicides to control the disease. There is also a significant current problem with leaf spot (*Alternaria dauci*) in carrots (T.J. Wicks, personal communication).

Sprouted seeds are commonly used as a vegetable in salads and cooking. Even low levels of seed contamination with fungi and bacteria cause major difficulties during sprouting. Mung bean (*Vigna mungo*) seed infection with *Macrophomina phaseolina* has caused major problems.

MAJOR ENDEMIC SEEDBORNE DISEASES IN NEW ZEALAND

Agricultural Crops

Pulses

Ascochyta lentis was first recorded in New Zealand in 1985 (Cromey *et al.*, 1987) and is now a common seedborne pathogen of lentil. Seed discoloration occurs where seed infection is severe. Seed treatments are frequently used to reduce disease levels in crops and a tolerant lentil variety has been released (Russell, 1994). Seedborne CMV was detected in lentil in experimental and commercial seed lines in New Zealand in 1994 with incidences of up to 2.5% being recorded (J.D. Fletcher, unpublished data).

Ascochyta blight (*Mycosphaerella pinodes* and *Phoma medicaginis* var. *pinodella*) is an important disease complex of peas in New Zealand. It affects yield, quality and market access of process and seed pea crops. Severe outbreaks occurred during 1993 and 1994 due to prolonged wet conditions. Disease surveys showed that the mean seedborne infection was 3.5% in 1993 and 8.5% in 1994, although infection levels up to 30% occurred in some crops. In 1995, the mean seedborne infection level declined to 0.6% (Grbavac, unpublished data). Seed treatment, voluntary testing of seed, crop rotation and crop hygiene are strategies for the controlling the disease.

Powdery mildew (*Erysiphe pisi*) is widespread in pea-growing areas, and may be seedborne (Crawford, 1927; Uppal *et al.*, 1935), although inoculum from volunteer plants or alternative hosts is more likely to instigate epidemics. During warm, dry seasons, this disease is very damaging in commercial and seed crops in New Zealand and Australia. Powdery mildew is controlled by the use of fungicides and resistant cultivars (Falloon *et al.*, 1993). Downy mildew (*Peronospora viciae*) of pea can be seedborne, although over-wintering oospores in plant debris are probably the more important source of inoculum.

Symptoms of seed-stain, tennis ball line and coat split attributed to PSbMV caused serious quality problems in pea seed during the mid-1980s. Seed infection levels up to 40% were recorded in field peas (Fletcher *et al.*, 1989). Control strategies include, education, voluntary testing and the incorporation of virus resistance (Russell and Jermyn, 1994).

Bacterial blight (*P. syringae* pv. *pisi*) has not been a problem on peas in New Zealand for the past 5 years. Voluntary testing of seed prior to planting followed by field inspection of crops has ensured that infected seed is not used for seed production. This control strategy has reduced the incidence of bacterial blight to an insignificant level. Field infection occasionally occurs as a result of the importation of pea seed from the USA and Europe which do not require testing prior to multiplication.

In navy beans, halo blight (*P. syringae* pv. *phaseolicola*) and bean common mosaic virus (BCMV) periodically cause concern in seed crops.

Cereals

Seed treatments are commonly used to control seedborne diseases of barley such as net blotch (*Drechslera teres*), spot blotch (*Drechslera sorokiniana*) and smuts (Close, 1983).

Pastures

Ustilago bullato causes head smut of many grasses, including prairie grass, an important pasture species in New Zealand. Falloon *et al.* (1988) reviewed the disease and methods of control. Head smut severely reduces seed herbage production and infected seed crops are rejected from certification. Crop health regulations and fungicide seed treatments are used to control this disease.

Seedborne AMV contributes to the widespread incidence of this disease in pastures and associated crops. Seed infection levels of 4% in lucerne (Forster *et al.*, 1985) and 13% in low-oestrogen red clover have been recorded (Fletcher, unpublished data).

Blind seed disease (*Gloeotinia granigena*) is a disease of ryegrass and fescue in Australia and New Zealand. The disease affects seed production, seed quality and market access. Disease epidemics occurred in ryegrass in Australia during 1987, 1988 and 1992 (Mebalds, 1993) and in New Zealand on ryegrass and fescue during 1992, 1993 and 1994 (Grbavac, unpublished data). Control of blind seed involves annual rotation of perennial ryegrass, use of urea fertilizer and systemic fungicides (Hampton and Scott, 1980, 1981; Mebalds, 1993). These strategies are very successful under low disease pressure, but are not effective when pressure is high. Recent research has shown that infected seed is the primary source of inoculum, and that urea reduces disease by increasing crop density which provides a barrier to ascospores preventing floral infection (Grbavac, unpublished data). Seed treatment is now recommended as the main strategy for control.

Horticultural Crops

Vegetatively propagated crops

Both Australia and New Zealand have similar disease problems on potatoes, hops and garlic. These are managed with the use of agrochemicals and 'high health' clonal material.

Diseases spread on true seed

Seedborne transmission of watermelon mosaic virus (WMV) and zucchini yellow mosaic virus (ZYMV) (1% and 7% respectively) in unharvested reject fruit appears to be significant for survival of these diseases in buttercup squash (*Cucurbita maxima*) (J.D. Fletcher, unpublished data). In New Zealand, diseases on brassica, carrot and onion seed are similar to those reported in Australia. Control is achieved using fungicide seed treatments.

IMPACT OF SEEDBORNE DISEASES ON TRADE

Seedborne diseases also have significant ramifications when selling seed both in domestic and international markets.

Domestic Marketing Requirements

Plant breeders and seed merchants have a 'duty of care' under Australian common law to ensure that seed is suitable for sale. They are liable for problems that arise from seed infection if they are negligent or mislead the purchaser of seed. Recent examples where this 'duty of care' has influenced the release of new cultivars and marketing of seed include PSbMV in field peas, CMV in lupins, and *A. lentis* and CMV in lentils.

International Market Requirements

Seed exports to overseas markets are often influenced by regulations governing diseases in seed. Australian lucerne (*Medicago sativa*) breeders have produced a number of varieties with international markets. Bacterial wilt (*Clavibacter michiganensis* subsp. *insidiosum*) is an important disease of lucerne (Stuteville and Erwin, 1990), but it has a restricted distribution and is not a problem within Australia. In many markets, declaration of area freedom is adequate; however, in some cases testing for seed infection is used and the pathogen is occasionally detected using sensitive tests. This variation in testing protocols affects markets where freedom from the pathogen is required. Similarly, testing for PSbMV and bacterial blight is routinely performed on pea seed exports where there is a market requirement.

ACKNOWLEDGEMENTS

We thank Rob O'Brien, Queensland Department of Primary Industries, and Dr Trevor Wicks, SARDI, for providing information about horticultural seedborne diseases in Australia.

REFERENCES

Ahmed, K.M. and Reddy, C.R. (1993) *A Pictorial Guide to the Identification of Seed-borne Fungi of Sorghum, Pearl Millet, Finger Millet, Chickpea, Pigeonpea and Groundnut.* International Crops Research Institute for the Semi-Arid Tropics (ICRISAT), Andhra Pradesh, India, 192 pp.

Alberts, E., Hannay, J. and Randles, J.W. (1985) An epidemic of cucumber mosaic virus in Australian lupins. *Australian Journal of Agricultural Research* 36, 267–273.

Alberts, E.V., Nutter, F.W. and Corbett, A.J. (1996) First report of alfalfa mosaic virus in lupins in Australia. *Plant Disease* 80, 1302.

Anonymous (1992) *Ergot of Sorghum.* Plant Quarantine Leaflet no 77. Australian Quarantine Inspection Service, Canberra, Australia.

Bertholini, S. (1995) Quality assurance and the law. In: *Proceedings of National Pulses Workshop: Pathology, Breeding and Agronomy.* South Australian Research and Development Institute, Adelaide, Australia, 54 pp.

Chandrashekar, M. and Culvenor, B. (1993) Exotic diseases of *Arachis, Cajanus, Cicer, Lens, Pisum, Vicia* and *Vigna.* Bureau of Resource Sciences Working Paper. Bureau of Resource Sciences, Canberra, Australia, 155 pp.

Close, R.C. (1983) Protecting the barley crop. In: Wright, G.M. and Wynn-Williams, R.B. (eds), *Barley: Production and Marketing*, pp. 73–78. Agronomy Society of New Zealand, Special Publication no. 2

Cooksley, D.A., Azad, H.R., Paulus, A.O. and Koike, S.T. (1991) Leaf spot of cilantro in California caused by a non-fluorescent *Pseudomonas syringae. Plant Disease* 75, 101.

Crawford, R.F. (1927) *Powdery Mildew of Peas.* New Mexico Experiment Station Bulletin no. 163, 13 pp.

Cromey, M.G., Mulholland, R.I., Russell, A.C. and Jermyn, W.A. (1987) *Ascochyta fabae* f.sp. *lentis* on lentil in New Zealand. *New Zealand Journal of Experimental Agriculture* 15, 235–238.

Dennis, J., Gooden, J. and Ramsey, M. (1995) Bacterial blight in coriander. In: *Proceedings of the 10th Biennial Australasian Plant Pathology Conference, Christchurch, New Zealand*, abstract 41, p. 44.

Diekmann, M. (1993) *Seed-borne Diseases in Seed Production.* International Centre for Agricultural Research in Dry Areas (ICARDA), Aleppo, Syria, 81 pp.

Falloon, R.E., Rolston, M.P. and Hume, D.E. (1988) Head smut of prairie grass and mountain brome; a review. *New Zealand Journal of Agricultural Science* 31, 459–466.

Falloon, R.E., Viljanen-Rollinson, S.L.H., McErlich, A.F., Scott, R.E., Goulden, D.S. and Bezar, H.J. (1993) Powdery mildew of peas: the costs and benefits of control. In: *Plant Protection: Costs, Benefits and Trade Implications.* New Zealand Plant Protection Society Inc., Christchurch, New Zealand, pp. 81–88.

Fletcher, J.D., Goulden, D.S., Russell, A.C. and Scott, R.E. (1989) Breeding for resistance to pea seed-borne mosaic virus. In: Hill, G.D. and Savage, G.P. (eds) *Grain Legumes: National Symposium and Workshop.* Agronomy Society of New Zealand, Special publication No 7, 124 (abstract).

Forster, R.L.S., Morris-Krisinich, B.A.M. and Musgrave, D.R. (1985) Incidence of alfalfa mosaic virus, lucerne Australian latent virus and lucerne transient streak virus in lucerne crops in the North Island of New Zealand. *New Zealand Journal of Agricultural Research* 28, 279–282.

Geering, A.D.W. (1992) The epidemiology of cucumber mosaic virus in narrow leafed lupins (*Lupinus angustifolius*) in South Australia. PhD Thesis, The University of Adelaide, Adelaide, Australia, 163 pp.

Gondran, J., Bournoville, R. and Duthion, C. (1994) *Identification of Diseases, Pests and Physical Constraints in White Lupin.* INRA/UNIP, Paris, France, 47 pp.

Gooden, J. and Alberts, E. (1995) A commercial test for bacterial blight in peas. In: *Proceedings of the 10th Biennial Australasian Plant Pathology Conference, Christchurch, New Zealand,* abstract 85, p. 57.

Gossen, B.D. and Morrall, R.A.A. (1981) The role of stubble and seed-borne inoculum in epidemic development of *Ascochyta* blight of lentils. *Canadian Journal of Plant Pathology* 3, 114.

Gossen, B.D. and Morrall, R.A.A. (1986) Transmission of *Ascochyta lentis* from infected seed and plant residue. *Canadian Journal of Plant Pathology* 8, 28–32.

Hampton, J.G. and Scott, D.J. (1980) Blind seed disease of ryegrass in New Zealand II. Nitrogen fertiliser: effect on indicence, and possible mode of action. *New Zealand Journal of Agricultural Research* 23, 149–153.

Hampton, J.G. and Scott, D.J. (1981) Blind seed disease in New Zealand. III. Urea: effect on apothecial production in the field and on blind seed infection – a note. *New Zealand Journal of Agricultural Research* 24, 233–234.

Haware, M.P., Nene, Y.L. and Mathur, S.B. (1986) *Seed-borne Diseases of Chickpea.* Technical Bulletin no. 1, Danish Government Institute of Seed Pathology for Developing Countries, Copenhagen, Denmark, 32 pp.

Kaiser, W.J. and Hannan, R.M. (1986) Incidence of seed-borne *Ascochyta lentis* in lentil germplasm. *Phytopathology* 76, 355–360.

Ketalaer, E., Diekmann, M. and Welzien, H.C. (1988) International spread of *Ascochyta rabiei* in chickpea seeds – an attempt at prognosis. *International Chickpea Newsletter* 18, 21–23.

Lazenby, A., Bartholomaeus, M., Boucher, B., Boyd, W.R., Campbell, A., Cracknell, R., Eagles, H., Lee, J., Lukey, G. and Marshall, B. (1994) *Trials and Errors: a Review of Variety Testing and Release Procedures in the Australian Grains Industry.* Grains Research and Development Corporation, Canberra, Australia, 224 pp.

Ligat, J.S., Cartwright, D. and Randles, J.W. (1991) Comparison of some pea seed-borne mosaic virus isolates and their detection by dot-immunobinding assay. *Australian Journal of Agricultural Research* 42, 441–51.

Mavridis, A., Meier zu Beerentrup, H. and Rudolph, K. (1989) Bacterial Umbel blight of Coriander in West Germany. In: Klement, Z. (ed.), *Plant Pathogenic Bacteria, Proceedings of the 7th International Conference on Plant Pathogenic Bacteria.* Academiai Kiadó, Budapest, Hungary, pp. 635–640.

Mebalds, K.I. (1993) The epidemiology and control of blind seed disease of perennial ryegrass. Master of Agricultural Science Thesis. School of Agriculture, La Trobe University, Victoria, Australia.

Nutter, F.W., Jones, R.A.C., Geering, A.D.W., Randles, J.W., Graetz, D. and Alberts, E.V. (1996) Quantification of alfalfa mosaic virus and cucumber mosaic virus epidemics in the lupin-medic pathosystem. *Phytopathology* 86, S14.

Persley, D. (1994) *Diseases of Vegetable Crops.* Queensland Department of Primary Industries, Brisbane, Australia, 75 pp.

Persley, D., O'Brien, R.G. and Syme, J.R. (1989) *Vegetable Crops – a Disease Management Guide.* Queensland Department of Primary Industries, Brisbane, Australia, 100 pp.

Russell, A.C. (1994) 'Rajah' lentil (*Lens cultinaris* Medik). *New Zealand Journal of Crop and Horticultural Science* 22, 469–470.

Russell, A.C. and Jermyn, W.A. (1994) 'Hadlee' field pea (*Pisum sativum* L.). *New Zealand Journal of Crop and Horticultural Science* 22, 221–222.

Stuteville, D.L. and Erwin, D.C. (1990) *Compendium of Alfalfa Diseases.* APS Press, St Paul, Minnesota, 84 pp.

Taylor, J.D. and Dudley, C.L. (1980) Bacterial disease of coriander. *Plant Pathology* 29, 117–121.

Toben, H.M. and Rudolf, K. (1996) *Pseudomonas syringae* pv. *coriandricola*, incident of bacterial umbel blight and seed decay in coriander (*Coriandrum sativum* L.) in Germany. *Journal of Phytopathology* 144, 169–178.

Uppal, B.N., Patel, M.K. and Karnat, M.N. (1935) *Pea Powdery Mildew in Bombay.* Bulletin no. 177, Department of Agriculture, Bombay, 12 pp.

van Vuurde, J.W.L. and van den Bovenkamp, G.W. (1989) Detection of *Pseudomonas syringae* pv. *phaseolicola* in beans. In: Saettler, A.W., Schraad, N.W. and Roth, D.A. (eds), *Detection of Bacteria in Seed and other Planting Material.* APS Press, St Paul, Minnesota, pp. 30–40.

Potentially Important Seedborne Pathogens of Economic Crops in North-east Asia

10

W.S. Wu

Department of Plant Pathology and Entomology, National Taiwan University, Taipei, Taiwan

INTRODUCTION

China, Korea, Japan and Taiwan are located in north-east Asia. The farming system in this region is diversification with a relatively small planting acreage. Most farmers use their own seed collected from their production fields. Some international seed companies have been established in this area for some time. Seedborne diseases are an important subject requiring intensive study mainly because they have been neglected or overlooked by local farmers, seed producers, and even governmental authorities. In this study, the important seedborne pathogens of the main groups of crops grown will be discussed. An important seedborne pathogen can be defined as a pathogen where seed is often the major source of inoculum and is its primary or important means of dispersal; the pathogen is often difficult to detect and identify, often has a severe impact on crop production and may be introduced via infected seed and difficult to control.

CEREALS

Rice is the major food crop in this region. Rice blast (*Pyricularia oryzae*) is a notorious disease and can be transmitted by seed (Richardson, 1979). The pathogen has been detected in Korea (Chung and Cho, 1981) and Japan (Kuniyasu, 1983), but not (recently) in Taiwan (Wu and Dow, 1993). Seed is not the only source of this disease (Ou, 1985). Brown spot and sheath blight,

(eds J.D. Hutchins and J.C. Reeves)

caused by *Bipolaris oryzae* and *Rhizoctonia solani*, respectively, are two other common and destructive diseases in this area.

Bacterial leaf blight caused by *Xanthomonas campestris* pv. *oryzae* is a prevalent disease in Korea (Chung and Cho, 1981), in the second rice planting season in Taiwan and has caused yield losses up to 50% in Japan (Ou, 1985). Seed transmission is not regarded as a major means of disease dispersal. The pathogen may survive in several weed hosts and rice stubble (Ou, 1985). The importance of seed transmission must therefore be clarified. *Burkolderia glumae* (formerly *Pseudomonas glumae*) causes grain rot, and glume blight is common in Japan (Goto and Ohata, 1956; Uematsu *et al.*, 1976) and the Philippines (Cottyn *et al.*, 1996a,b).

Wheat, barley, maize and sorghum are important crops in northern China, Korea and Japan. Loose smut of wheat and barley is a severe problem and has become less important since it can be effectively controlled by hot water treatment and systemic fungicides. Rusts, southern corn leaf blight and stalk rot or ear and kernel rots are the most prominent diseases among maize fields (Chung and Cho, 1981; Li and Wu, 1986). Stalk and kernel rot caused by *Fusarium moniliforme* is common in Taiwan (Li and Wu, 1986). The presence of this pathogen on maize seed is common. Sorghum is an important crop in northern China. *Fusarium moniliforme* is the most common seedborne pathogen on sorghum seed (Wu, 1983).

VEGETABLES

The varieties of vegetables grown in this region are numerous. Alliaceous, cruciferous, cucurbitaceous, leguminous, solanaceous and umbelliferous crops are frequently planted.

Alliaceous Crops

Purple spot caused by *Stemphylium vesicarium* commonly exists on onions, Welsh onions and garlic. *Stemphylium botryosum* predominates on onion seeds (Wu, 1977, 1979; Neergaard, 1979) and causes damping off of onion seedlings.

Cruciferous Crops

Alternaria brassicicola is widespread on various cruciferous crops, e.g. broccoli, cabbage, cauliflower, radishes and turnips (Wu, 1979; Wu *et al.*, 1979; Tohyama and Tsuda, 1995). This pathogen exists in all major production fields of cruciferous crops globally (Rotem, 1994). Effective seed treatment with antagonists and chemicals to control this pathogen has been achieved (Wu and Lu, 1984). Black rot is another common disease among cruciferous crops in fields. No research appears to indicate that the black rot observed in the field is caused by seedborne *Xanthomonas campestris* pv. *campestris* in this region.

Cucurbitaceous Crops

There are many important cucurbitaceous vegetables and fruits, e.g. balsam pear, bottle gourd, cantaloupe, Chinese okra, cucumber, muskmelon, oriental pickling melon, snake gourd, sponge gourd, squash, vegetable pear, watermelon, wax gourd. Fruits of many of these crops lie on the ground. Consequently, soilborne plant pathogens may infect and penetrate fruits and the seeds may become infected. Species of *Fusarium* have been isolated from pumpkin seeds (Tezuka *et al.*, 1993), seeds of bottle gourd (Kuniyasu, 1983) and on cucurbit seeds in Taiwan (Hwang and Wang, 1995). These fungi are believed to be seedborne pathogens, but their occurrence on seeds and their pathogenicity has not been determined. *Colletotrichum lagenarium* is in the list of important pathogens regulated in Korea (Chung and Cho, 1981). This pathogen commonly occurs in various cucurbit fields in Taiwan. The most susceptible crops are watermelon, cucumber and muskmelon (Hsieh, 1994). However, no study has been made regarding the effects of the pathogen on emergence and yield.

Six different viruses, namely zucchini yellow mosaic virus (ZYMV), papaya ringspot virus type W (PRV-W), cucumber mosaic virus (CMV), cucumber green mottle mosaic virus (CGMMV), melon vein-banding mosaic virus (MVbMV) and tomato spotted wilt virus (TSWV) are known to be present in cucurbit fields in Taiwan (F.Y. Wu *et al.*, 1994). Of these six viruses, CMV, CGMMV and TSWV are seedborne. Bottle gourd is commonly infected with CGMMV. CMV is frequently isolated from sponge gourd and muskmelon. TSWV causes damage in watermelon and muskmelon fields. The amount of disease caused by these pathogens in Taiwan has not been quantified. CGMMV has caused serious damage to cucumber in south-west Japan since 1966 (Kuniyasu, 1983), and watermelons have been seriously infected by CGMMV in the eastern central part of Japan since 1968.

Leguminous Crops

Many economically important crops belonging to the *Leguminosae*, including adzuki bean, asparagus bean, broad bean, kidney bean, lima bean, mung bean, pea, peanut, soybean and yam bean, are commonly planted in this region. Few studies on seedborne diseases of these crops have taken place.

A destructive wilt of adzuki bean occurred in Hokkaido, Japan, in 1983 and the total acreage of diseased plants was about 2000 ha in 1985 (Kondo, 1995) and was believed to have been caused by *Fusarium oxysporum* f. sp. *adzukicola*, which is frequently isolated from seeds. Bean common mosaic virus (BCMV), blackeye cowpea mosaic virus (BlCMV), cucumber mosaic virus (CMV) and peanut stripe virus (PStV) have been detected in adzuki fields in Taiwan (Hsu *et al.*, 1987; Chang and Chen, 1990; Chang *et al.*, 1990). Asparagus beans are particularly susceptible to virus infection. These beans are frequently infected with both CMV and BlCMV resulting in severe rugose mosaic (Chang, 1983). The yield from plants grown from healthy seed is 11–74% higher than that from diseased plants (Chang *et al.*, 1994).

Leaf blight of pea occurs severely in Taiwan and is identified as being caused by seedborne *Mycosphaerella pinodes* (Chen and Huang, 1994; Chen *et al.*, 1994). Besides this important fungal pathogen, pea seedborne mosaic virus is widespread in pea fields in Taiwan (Chang *et al.*, 1989). Soybean purple blotch caused by *Cercospora kikuchii* is regarded as the most important pathogen needed to be inspected (Chung and Cho, 1981). The emergence of infected seed is reduced by 45% in the field compared with healthy seed (Oh and Kwon, 1981). Seed infection rates can occasionally exceed 60% (Kuniyasu, 1983). Pod and stem blight caused by *Phomopsis sojae* has been found in Taiwan (Wu and Lee, 1984) and Korea (Park *et al.*, 1992). This disease directly affects the quality and quantity of soybean. Soybean mosaic virus (SMV) has also been detected from several seed lots (La and Cho, 1981).

Solanaceous Crops

Peppers and tomatoes are the two most important vegetables cultivated in this area. Hot and sweet peppers have many common important diseases. Anthracnose may be caused by *Colletotrichum gloeosporioides*, *C. capsici*, *C. dematium*, *C. acutatum* and *C. coccodes* (Black *et al.*, 1991). The pathogen penetrates the fruit and the seeds become readily infected. No pepper line is immune to this disease. Seed health tests are required before sowing a crop.

Early blight and *Fusarium* wilt are two common diseases of tomatoes caused by *F. oxysporum* f.sp. *lycopersici*. Inoculum of these pathogens occurs in seed and from other sources (e.g., weeds, soil and debris). The pathogens of these two diseases have recently been undetected from seeds of solanaceous crops in Taiwan (Chen and Wu, 1995). No recommended methods are available for detecting these important seedborne pathogens. Tomato bacterial canker caused by *Clavibacter michiganensis* subsp. *michiganensis* is thought to have been introduced into Japan through imported infected seed since 1958 (Kuniyasu, 1983). Since then, this disease has resulted in significant damage to Japanese tomato crops.

Umbelliferous Crops

Carrot and celery are important umbelliferous vegetables. Both of them prefer to grow in cool seasons. Seeds of those crops, which are planted in Taiwan and Korea, are imported primarily from Japan. *Alternaria radicina* and *Alternaria dauci* have been detected in carrot and celery seeds and initiated severe disease epidemics (Lo, 1973; Chen and Wu, 1995). *Alternaria dauci* has been detected on carrot seed in Korea (Cho *et al.*, 1988).

Miscellaneous Crops

The important seed-transmitted disease of spinach is downy mildew. *Peronospora effusa* has been determined to be transmitted by seeds (Inaba *et al.*, 1983).

The percentage of infected seedlings is positively correlated with the amount of oospore-infested seed sown. Downy mildew of spinach is regarded as a serious problem in Japan (H. Shiomi, personal communication).

ORNAMENTAL CROPS

Few major commercial seed companies in this region produce ornamental seeds; most ornamental plant seed is imported from other regions. *Alternaria zinnia* and *A. carthami* have been detected from imported zinnia seeds in Taiwan (Wu and Yang, 1992; Chou and Wu, 1995) and Korea (Yu and Park, 1988). *A. tagetica* is thought to have been introduced into Korea through imported infected marigold seeds (Yu and Lee, 1989). However, no data are available to quantify the losses of the relative crops caused by these pathogens. Cho *et al.* (1988) describe several occurrences of imported infected seed introducing a pathogen to an otherwise uninfected area.

DISCUSSION

Accumulating information regarding seedborne pathogens in north-east Asia is a relatively difficult task, possibly due to the potential threat from seedborne diseases being overlooked. Of particular importance are seedborne virus and bacterial diseases which cause severe disease losses in the field (e.g. blackeye cowpea mosaic virus and pea seedborne mosaic virus). Methods for differentiating between important seedborne pathogenic fungi including species of *Fusarium* and *Alternaria* are urgently required. Attempts must also be made to test for pathogens which are not endemic to the region but may pose a threat from the import of infected seed (e.g. ornamental plants and turf grasses).

REFERENCES

Black, L.L., Green, S.K., Hartman, G.L. and Poulos, J.M. (1991) *Pepper Diseases, a Field Guide.* Publication no. 91–347, AVRDC, Taiwan, 98 pp.

Chang, C.A. (1983) Rugose mosaic of asparagus bean caused by dual infection with cucumber mosaic virus and blackeye cowpea mosaic virus. *Plant Protection Bulletin.* 25, 177–190. [In Chinese, with English summary.]

Chang, C.A. and Chen, H.C. (1990) Survey and identification of adzuki bean viruses occurring in southern Taiwan. *Plant Protection Bulletin*, 32, 331 [Abstract].

Chang, C.A., Chen, C.M., Chen, H.C., Deng T.C. and Huang, C.H. (1989) Pea seed-borne mosaic virus causing pea mottling in Taiwan. *Plant Protection Bulletin* 31, 366–376.

Chang, C.A., Purcifull, D.E. and Zettler, F.W. (1990) Comparison of two strains of peanut stripe virus in Taiwan. *Plant Disease* 74, 593–596.

Chang, C.A., Yang, T.T., Tsan, T.M. and Chen, C.C. (1994) Production and application of virus free seeds to control virus disease of asparagus beans. *Plant Protection Bulletin.* 36, 313–325. [In Chinese, with English summary.]

Chen, M.H. and Huang, J.W. (1994) Factors affecting seed transmission of leaf blight pathogen of garden peas, *Mycosphaerella pinodes. Plant Protection Bulletin* 36, 189–200. [In Chinese, with English summary.]

Chen, M.H., Huang, J.W. and Yein, C.F. (1994) The pathway and influence factors in pod and seed infections of garden peas by *Mycosphaerella pinodes. Plant Pathology Bulletin.* 3, 133–139. [In Chinese, with English summary.]

Chen, T.W. and Wu, W.S. (1995) Detection of seed-borne fungi of vegetables in Taiwan. *Plant Pathology Bulletin* 4, 17–24.

Cho, S.H., Kang, H.D. and Kim, C.H. (1988) Fungi associated with foreign seeds of twenty three plant species from six countries. *Korean Journal of Plant Pathology* 44, 319–324.

Chou, J.K. and Wu, W.S. (1995) Seed-borne fungal pathogens of ornamental flowering plants. *Seed Science and Technology* 23, 201–209.

Chung, H.S. and Cho, N.G. (1981) Present status and problems of seed health testing in Korea. In: *Proceedings of Seed Pathology Workshop, Seoul National University, Suweon, Korea*, pp. 8–49.

Cottyn, B., Cerez, M.T., Van Outryve, M.F., Barroga, J., Swings, J. and Mew, T.W. (1996a) Bacterial diseases of rice. I. Pathogenic bacteria associated with sheath rot complex and grain discoloration of rice in the Philippines. *Plant Disease* 80, 429–437.

Cottyn, B., Van Outryve, M.F., Cerez, M.T., DeCleene, M., Swings, J. and Mew, T.W. (1996b) Bacterial diseases of rice. II. Characterization of pathogenic bacteria associated with sheath rot complex and grain discoloration in the Philippines. *Plant Disease* 80, 438–445.

Goto, K. and Ohata, K. (1956) New bacterial diseases of rice brown stripe and grain rot. *Annals of the Phytopathological Society of Japan* 21, 46–47.

Hsieh, C.P.Y. (1994) Anthracnose of cucurbits. In: *Proceedings of Symposium on Techniques of Cucurbit Protection.* Plant Protection Society of ROC, Taiwan, pp. 115–125.

Hsu, S.H., Huang, C.H. and Chang, C.A. (1987) Identification of bean common mosaic virus isolated from adzuki beans. *Plant Protection Bulletin* 29, 447. [Abstract in Chinese.]

Hwang, C.W. and Wang, L.M. (1995) Pathogenicity, survival and control of *Fusarium* wilt of cucubits. In: *Proceedings of Symposium on Techniques of Cucurbit Protection.* Plant Protection Society of ROC, Taiwan, pp. 127–134.

Inaba, T., Takahashi, K. and Morinaka, T. (1983) Seed transmission of spinach downy mildew. *Plant Disease* 67, 1139–1141.

Kondo, N. (1995) Studies on adzuki bean wilt caused by *Fusarium oxysporum. Bulletin of Agricultural Faculty, Hokkaido University* 195, 441–472. [In Japanese, with English summary.]

Kuniyasu, K. (1983) The present situation of seed-borne diseases and seed disinfection in Japan. *Japan Pesticide Information* 43, 13–22.

La, Y.J. and Cho, Y.S. (1981) Seed-borne diseases of plant caused by viruses and bacteria, and their control in Korea. In: *Proceedings of Seed Pathology Workshop, Seoul National University, Suweon, Korea*, pp. 74–77.

Li, S.Y. and Wu, W.S. (1986) Significance and control of seed-borne *Fusarium moniliforme* of corn. *Plant Protection Bulletin* 28, 191–202.

Neergaard, P. (1979) *Seed Pathology*, 2 vols. MacMillan Press Ltd, London, 1191 pp.

Oh, J.H. and Kwon, S.H. (1981) Evaluation of native soybean collection for resistance to purple seed stain. In: *Proceedings of Seed Pathology Workshop, Seoul National University, Suweon, Korea*, pp. 114–120.

Ou, S.H. (1985) *Rice Diseases*. CAB International, Mycological Institute, Kew, UK, 380 pp.

Park, E.W., Lee, C.S. and Hong, E.H. (1992) Forecasting *Phomopsis* seed decay by the soybean pod infection test. *Korean Journal of Plant Pathology* 82, 96–100.

Richardson, M.J. (1979) An annotated list of seed-borne disease. CMI, Kew, 320 pp.

Rotem, J. (1994) *The Genus Alternaria, Biology, Epidemiology, and Pathogenicity*. APS Press, St Paul, Minnesota, 326 pp.

Tezuka, N., Horiuchi, S. and Takeuchi, S. (1993) Isolation of *Fusarium solani* from pumpkin seeds and the pathogenicity. *Proceedings of Kansai Plant Protection Society* 35, 87–88. [In Japanese.]

Tohyama, A. and Tsuda, M. (1995) Alternaria on cruciferous plants. 4. *Alternaria* species on seed of some cruciferous crops and their pathogenicity. *Mycoscience* 36, 257–261.

Uematsa, T., Yoshimura, D., Nishiyama, K., Ibaraki, T. and Fujii, H. (1976) Pathogenic bacterium causing seedling rot of rice. *Annals of the Phytopathological Society of Japan* 42, 464–471.

Wu, F.Y., Hseu, S.H. and Huang, C.H. (1994) Viruses of cucurbit crops in Taiwan. In: *Proceeding of Symposium on Techniques of Cucurbit Protection*. Plant Protection Society of ROC, Taiwan, pp. 159–168. [In Chinese, with English summary.]

Wu, W.S. (1977) Black mold rot of onion in Taiwan. *Plant Protection Bulletin* 19, 202–205.

Wu, W.S. (1979) Survey on seed-borne fungi of vegetables. *Plant Protection Bulletin* 21, 206–219.

Wu, W.S. (1983) Sorghum diseases in Taiwan and characterization and control of its seed-borne pathogens. *Plant Protection Bulletin* 25, 1–13.

Wu, W.S. and Dow, S.K. (1993) A survey of rice seed-borne fungi in Taiwan. *Plant Pathology Bulletin* 2, 52–55.

Wu, W.S. and Lee, M.C. (1984) Occurrence, pathogenicity and control of *Phomopsis sojae* on soybean. *Memoirs of the College of Agriculture NTU* 242, 16–26.

Wu, W.S. and Lu, J.H. (1984) Seed treatment with antagonists and chemicals to control *Alternaria brassicicola*. *Seed Science and Technology* 12, 851–862.

Wu, W.S. and Yang Y.H. (1992) Alternaria blight, a seed-transmitted disease of zinnia in Taiwan. *Plant Pathology Bulletin* 1, 115–123.

Wu, W.S., Wu, C.H.H. and Wu, K.C. (1979) *Alternaria brassicicola*, a destructive pathogen in seed production of cauliflower. *Plant Protection Bulletin* 21, 294–304.

Yu, S.H. and Park J.S. (1988) Leaf spot of zinnia caused by *Alternaria zinniae* M.B. Ellis in Korea. *Korean Journal of Plant Pathology* 42, 85–87.

ISTA-PDC's Terms of Reference in the Past and for the Future

11

C.J. Langerak

DLO Centre for Plant Breeding and Reproduction Research (CPRO-DLO), Wageningen, The Netherlands

ISTA-PDC, CONSTITUTION AND MEMBERSHIP

The Plant Disease Committee (PDC) is one of the 17 Technical Committees of the International Seed Testing Association (ISTA). It consists of a maximum of 13 members presided over by a Chairman and Vice-chairman, all officially appointed by the Executive Committee of the ISTA. A variable number of persons may have the status of co-leader and assist an appointed member with specific tasks. The constitution of the PDC is reconsidered every 3 years following an ISTA Congress. Contributions of members have always been given on a voluntary basis, which is also the case for any other person involved in PDC activities.

Until 1987, the year the 19th PDC Seminar was organized in Wageningen, The Netherlands, all members of the PDC fulfilled jobs in governmental services such as official seed testing stations, plant protection services, universities or research institutes. Such positions automatically implied involvement in seed health testing for certification, research, legislation matters, teaching or training. Due to such a wide diversity of professional tasks, a broad insight has been guaranteed into the problems seedborne diseases can cause in agriculture, horticulture and in the seed trade. Representatives of private seed companies also participated in this seminar. This initiated the more pronounced and direct involvement of the private seed industry in the PDC programme. Consequently, a representative from this area has been appointed as a PDC member since the last ISTA Congress in 1995, Copenhagen, Denmark.

TERMS OF REFERENCE IN THE PAST AND ACTIVITIES

A historical overview of the activities of the ISTA-PDC has been presented previously (Langerak, 1993). In that presentation it was stated that the main goal of the PDC is the 'international standardization of seed health testing methods' which should result in the lowest possible variation in test results within and between laboratories. Other terms of reference deal with the dissemination of knowledge and experience. This includes production of handbooks and organization of meetings and training courses.

Standardization of Methods

Methods are standardized by PDC Working Groups comprising members experienced with the pathogen and the method under examination. Working Group members participate in a comparative test and may attend a technical workshop where aspects of a test are demonstrated and critically discussed. Traditionally the work focused on seedborne fungi and was carried out by groups with specialization in groups of crops, e.g. vegetables, temperate cereals and grasses, cruciferous crops, etc. In the late 1970s, two new Working Groups were established and charged with the development of detection methods for seedborne bacteria and viruses. The work in these two groups was directed to sub-Working Groups, each covering the methodology on one host–pathogen combination or studies on a specfic subject related to the activities of the Working Group. A similar concept was adopted for the work on seedborne fungi after the ISTA Congress in 1992, Buenos Aires, Argentina.

The greatest output in the field of method standardization was reached in the period 1975–1987. This was also a period in which new techniques based on immunology were introduced, e.g. enzyme-linked immunosorbent assay (ELISA) and immunofluorescence microscopy. These techniques proved to be more rapid methods for the detection of seedborne viruses and bacteria compared with traditional growing-on and dilution plating tests. Though these new techniques looked promising for introduction as routine seed health testing methods, progress in international standardization was rather slow due to poor availability of specific high quality antisera and differences in interpretation of method descriptions.

Simultaneously, detection methods for seedborne fungi were standardized further. As a result a number of new Working Sheets were incorporated in the *Handbook on Seed Health Testing*, which comprises an 'Annotated List of Seedborne Diseases' (section 1.1), an Introduction to Methods of Seed Health Testing' (section 2.1), and Working Sheets, each dealing with method(s) for detection of one pathogen on one host (section 2.2). Furthermore, detection methods for 14 seedborne fungus–host combinations have been incorporated in the ISTA Rules as recommended methods.

PDC Workshops, Seminars and Symposia

Exchange of knowledge has traditionally been established through PDC workshops called PDC seminars (since 1981) and PDC symposia (since 1993). The aims of the workshops, seminars and symposia have changed over the years. The workshops and seminars brought PDC Working Group members together to evaluate test methods in the laboratory, with a view to agreeing on methods that might be adopted internationally and, presumably, eventually included in the ISTA Rules. In the symposia organized in Ottawa (1993) and Cambridge (1996), these technical elements were omitted from the programme and directed to specific technical workshops. The 'Serology' Bacteriology sub-Working Group organized a technical workshop in 1987, Wageningen, The Netherlands in order to standardize the use of specific immunological techniques. Attention was paid to specificity testing and standardization of antisera, an aspect which was – and still is – often underestimated. Centralizing antiserum production and testing internationally was recommended in the report of this sub-working group (Anon., 1988). This idea, however, has not been realized until recently (see Chapter 32). PDC symposia, organized every 3 years (since 1993), aim to bring together a wide range of scientists working on seedborne disease problems. The concept is to provide a forum for the exchange of information on seed health testing and related topics.

PDC Training Courses

Training courses are organized under the supervision of the Pathogen Working Groups. Experienced PDC members demonstrate test methods to inexperienced seed pathologists. The aim of such courses is to teach and to train participants in the theoretical and technical aspects of seed health testing, thereby following the standardized methodology recommended by ISTA-PDC Working Groups. These courses are specifically organized for areas where there is an interest in seed health testing, but where a programme of testing does not exist.

RECENT DEVELOPMENTS AFFECTING TERMS OF REFERENCE AND PROGRESS

During the last decade significant scientific, political and economical changes on a worldwide basis have influenced the PDC's activities and progress, and might require adaptation of its terms of reference; these changes will be discussed in this chapter.

Advances in Knowledge about Seed Pathology

Over the years there has been a considerable increase in the field of seed pathology demonstrated by the increasing numbers of participants, both in

PDC seminars and symposia and in similar meetings on other occasions. Unfortunately this disturbed progress in standardization of seed health testing methods in the late 1980s. Too many applicants for the general PDC workshop/seminars, which covered the bench work, made continuation of this kind of general meeting impossible. Splitting up these meetings into 'pathogen-type' workshops was the only solution. By doing that, the regular contact between leading PDC members and contributors faded away. Consequently, the process of method standardization slowed down. Neither were new criteria formulated for incorporation of newly standardized methods into Working Sheets and Rules. A solution for this problem was sought by the reorganization of these PDC seminars.

A quite different aspect of a worldwide advanced knowledge on seedborne pathogens relates to the way the economical importance of these pathogens is judged on the basis of the ISTA-PDC's 'Annotated List of Seedborne Diseases' (Richardson, 1990). This list acts as a comprehensive source of information describing microorganisms that have been found on seeds of plant species of agricultural and horticultural importance, and to a lesser extent, in forestry. However, the latest edition does not give an indication of the economic importance of these microorganisms. A more critical attitude is required for a correct and relevant interpretation and use of the information provided. Incorrect interpretation of the information has often resulted in unjustified barriers to trade. Listed organisms should not be arbitarily placed on a quarantine list or a national list with importation requirements without consultation with experts and an assessment of the risk that such organisms pose.

Recent developments in molecular biology offer new tools which may be useful in seed health testing. Several variants of polymerase chain reaction (PCR) techniques (Langerak *et al.*, 1996; see also Section III, this volume) describe new approaches for the detection of seedborne bacteria, fungi and viruses. It should be realized that with these techniques it is still difficult to quantify infection levels and no direct interpretation can be given about the viability and/or pathogenicity of the detected pathogen. A technical workshop may act as the most suitable forum to discuss the interpretation of results obtained using these techniques.

Financial and Economic Developments

The last 10 years have been characterized by significant changes in the status of many public services. Privatization of public services, including those charged with domestic seed production and seed certification, has undoubtedly influenced the progress of the PDC. Many PDC members have found it increasingly difficult to contribute on a voluntary basis to PDC activities. It is not surprising that the seed industry has developed various initiatives to cooperate, both nationally and internationally, to continue with the standardization of seed health testing methods.

Further globalization of the seed industry is expected in the near future. Concurrent with this trade is the increased risk of spread of specific seedborne pathogens to areas where they have not previously been detected. Such globalization could not be overlooked, and therefore the PDC has broadened the team of members and co-leaders in the committee aiming to enhance the exchange of knowledge and experiences.

PROSPECTS

Following the establishment of the International Seed Health Initiative (ISHI) on seedborne diseases (Meijerink and Van Breukelen, 1995; see also Chapter 12) both parties have agreed to work together to standardize a protocol for comparative testing programmes. These guidelines are now in preparation and will be based on ISO Guide 43, *Proficiency Testing by Interlaboratory Comparisons*, and *Guidelines for Collaborative Study Procedures to Validate Characteristics of a Method of Analysis* (J.W. Sheppard, personal communication) which has been derived from a similar document used by the Association of Official Analytical Chemists (Anon., 1989).

Quality assurance has become a new aspect in the conduct of seed health tests. The integration of new technologies into routine use will require the availability of (certified) standardized antisera, DNA probes and microbiological reference materials. It should therefore be an important goal for each Pathogen Working Group to indicate the quality of antisera and reference materials they require for performance of reliable tests. The Virology Working Group has made progress in this direction. Establishment of a serum bank where these antisera will be available is seen as a prerequisite for developing further activities in this Working Group (see Chapter 32). The main problem encountered is finding the finance to initiate these research programmes. The ISTA-PDC has no such resources for funding such projects and therefore should contribute more actively in supporting and recommending project proposals. The PDC recently encouraged and supported a project to develop reference materials for use in detecting seedborne bacteria. This proposal was submitted to the FAIR programme of the European Commission and the project commenced in 1996 (see Chapter 31).

To what extent should the PDC deal with the establishment of threshold levels for disease transmission via seeds when its main term of reference covers the standardization of seed health testing methods? The PDC could produce guidelines concerning the economically important seedborne pathogens on a worldwide basis. These guidelines would integrate information on tolerance levels of seed infection, disease epidemiology, seed sampling and testing procedures and seed treatment. This topic will be addressed in the next ISTA-PDC symposium (to be held in 1999) and should provide a sound basis for improvements in seed health into the 21st century.

REFERENCES

Anonymous (1988) Further activities and discussions in the serology subgroup. In: *Report of the 1st International Serology Workshop*, pp. 39–40.

Anonymous (1989) Guidelines for collaborative study procedures. *Journal of the Association of Official Analytical Chemists* 72, 694–704.

Langerak, C.J. (1993) '75 years of Seed Pathology', past, present and future prospects. In: Sheppard, J.W. (ed.), *Proceedings of the First ISTA Plant Disease Committee Symposium, Ottawa, Canada*, pp. 1–2.

Langerak, C.J., Van den Bulk, R.W. and Franken, A.A.J.M. (1996) Indexing for pathogens. In: De Boer, S.H., Tommerup, I.C. and Andrews, J.H. (eds) *Advances in Botanical Research*, Vol. 23, *Pathogen Indexing Technologies.* Academic Press, London, pp. 171–215.

Meijerink, G.A.A.M. and Van Breukelen, E.W.M. (1995) Seedborne diseases: International Initiative standardizes test protocols. *Prophyta Annual* 49, 58–65.

Richardson, M.J. (1990) An annotated list of seedborne diseases. In: *International Seed Testing Association Handbook on Seed Health Testing*. ISTA, Zürich, Switzerland.

The International Seed Health Initiative

G. Meijerink

S+G Seeds BV, Westeinde 62, PO Box 26, 1600 AA Enkhuizen, The Netherlands

INTRODUCTION

Vegetable growers around the world are confronted with the increasingly difficult task of having to produce high quality products coupled with an increasing vulnerability due to the risk of the outbreak of a disease epidemic. Some of the factors that lead to this increased risk include the use of transplants and irrigation, the shift of vegetable production from one country or region to another, the trend towards monoculture and the larger scale of production.

Financial consequences of disease epidemics are enormous in modern vegetable production, where large amounts of money per hectare are invested by growers. Recent incidents in various countries have demonstrated that multi-million losses may occur. They include: tobamovirus in pepper crops in Spain (1991–1992); watermelon fruit blotch in the southern states of the USA (1993–1995); and *Clavibacter michiganensis* in glasshouse tomato crops in Belgium (1995–1996). These incidents have led to financial claims and court cases against plant growers and seed suppliers. In addition to direct financial consequences these incidents also have negative consequences for the image of the seed/plant supply chain, as they attract a lot of publicity.

Healthy seed is seen as a prerequisite for success in efficient crop production. It is logical that growers are requesting more assurance regarding the supply of healthy vegetable seed. The international seed industry is very willing to meet these requests, but achieving this presents a number of difficulties.

THE CURRENT SITUATION

The vegetable seed industry is an international business: seeds of a specific variety may be produced in east Asia, cleaned, treated and packed in Western Europe, sold to the USA, to be finally distributed and planted in a South American country. Production and distribution of healthy seed is therefore an international issue, where international agreements on seed health management methods and standards, based on sound scientific and practical technology, are required. National and international plant protection and quarantine offices and associations, such as the International Seed Testing Association (ISTA), have since long recognized this need.

As grower requirements for healthy seeds have increased, these efforts need to be intensified. Growers do not feel secure if seed suppliers simply meet the (inter)national regulations, they require more assurance and guarantees from their suppliers. Research at institutes, universities and seed companies, has resulted in a variety of assay methods and epidemiological thresholds for several pathogens. This only confuses both the growers and seed companies.

There is a lack of uniformity in test methods and protocols which results in differences depending upon which country or seed company has conducted the test, e.g. lettuce mosaic virus ELISA assay on 2300 seedlings versus 30,000 seeds, tobamovirus ELISA assay versus bioassay and *Xanthomonas campestris* assays on sample sizes ranging from 10,000 to 100,000 seeds. This leads to confusion and inefficiency, whereby seeds already tested in one country/region may have to be retested for importation or use in another country/region.

Slow development of new test methodology for some pathogens results in the continued use of laborious methods, heavily dependent upon individual analyst skills and with limited reproducibility between laboratories e.g. *Phoma lingam* assay in Brassica seed and *Phoma apiicola* assay in celery. Methods adapted to measure the efficacy of treatment of commercial vegetable seed lots are often not available or of dubious quality. Efficiency of seed treatment in relation to the initial infection level is often not known, e.g. in the case of sodium hypochlorite treated tomato seeds (*Clavibacter michiganensis*) and carbendazim treated Brassica seeds (*Phoma lingam*). This means that it is impossible to give vegetable growers the 'firm guarantees' that they ask for. It also means that seed companies lack reliable and recognized methods for checking the health status of seed lots and for making the correct seed health management decisions. Absence of these assays and thresholds is missed by all parties in litigation suits where it is unclear where the responsibilities of the seed supplier and grower lie.

The international vegetable seed industry is of the opinion that more effort is required to achieve a new broadly supported balance between market requirements and seed health specifications. This need is such that seed companies of five major seed producing countries, representing more than 70% of the world trade volume in vegetable seeds, have agreed to overcome competitive differences and to cooperate in the area of seed health management. This resulted in the setting up of an International Seed Health Initiative (ISHI) in 1994.

THE INTERNATIONAL SEED HEALTH INITIATIVE

The seed companies currently cooperating in the ISHI are from France, Israel, Japan, The Netherlands and the USA. These companies have a broad practical experience in testing hundreds of seed lots per year and they have information about the practical performance of these lots in different countries and climatic regions. From the start of the ISHI these companies have cooperated with inspection services and institutes, such as the Dutch Inspection Service for Vegetable and Flower seeds (NAKG), the French National Station for Seed Testing (SNES), The Dutch Research Centre for Plant Production and Reproduction (DLO-CPRO) and the Seed Science Centre (SSC) of Iowa State University. Private laboratories such as Seed Testing of America (STA) have also participated. Cooperation with other institutes and laboratories is foreseen. Combining and utilizing this knowledge and experience is the major aim of the ISHI.

The mission statement of the ISHI has the following objectives: (i) to secure the delivery of sufficiently healthy seeds to our customers on a worldwide basis; (ii) to assess and develop suitable test protocols, establish adequate pathogen thresholds and develop an information database for the control of seedborne pathogens of international economic importance for vegetable crops; and (iii) to seek recognition and cooperate with official national and international regulatory and accreditation authorities. These protocols, thresholds and other measures will enhance customer satisfaction, facilitate the international movement of seeds, support the international seed industry in improving product quality and assist seed companies in risk management.

THE STEPS AND ACHIEVEMENTS OF THE ISHI

The structure of ISHI was defined in 1994–1995 and includes three levels of organization, as outlined below.

Policy Coordination Group

The Policy Coordination Group (PCG) consists of management representatives from each company (one per country), assisted by the chairman of the Technical Coordination Group (TCG). This Group meets twice a year and: (i) defines the objectives of the initiative; (ii) coordinates the non-technical issues (such as communication and promotion of the initiative, financing of activities and contracts for joint research projects); and (iii) handles exchange of knowledge between participants.

Technical Coordination Group

The TCG consists of members of the Technical Groups for each species, country representatives and representatives of the participating institutes and inspection services. This Group meets once or twice a year and has the following main tasks: (i) organizing International Technical Groups (ITGs) for each species; (ii) setting up the system of plant pathogen surveys and comparative testing; (iii) coordinating progress evaluation, planning meetings, joint research projects and workshops; and (iv) coordinating cooperation with the ISTA Plant Disease Committee (PDC).

International Technical Groups

The ITGs consist of seed industry experts for each species (one per country). The Groups form the heart of the ISHI. Currently there are seven ITGs for the following species and plant families (beans, brassicas, carrot, cucurbits, lettuce, pepper, tomato). Their main tasks are: (i) to coordinate the technical issues at the species/plant family level (preparation and updating of crop/pathogen surveys, defining objectives for comparative testing, evaluating comparative test plans, validation of comparative test reports, proposal of follow-up of comparative tests and assessment of training requirements); and (ii) to liaise with the ISTA-PDC Working Groups. Most country experts are backed by national working groups.

Additionally there is a separate structure for coordinating comparative tests with the NAKG serving as the central coordinator. Currently the NAKG, SNES and STA coordinate and supervise comparative tests allowing the ISHI to perform these tests in a reliable and cost-efficient way.

Prioritization of Vegetable Species and Pathogens

With more than 20 participating seed companies from five different countries, commercially breeding over 60 different vegetable species, for which more than 200 different seedborne pathogens could be identified, it was necessary to focus and agree on a limited number of species and pathogens to work on. The outcome of this prioritization exercise is shown in Table 12.1.

Surveying and Sharing of Available Knowledge of Priority Pathogens

A survey/questionnaire was developed in order to obtain a comprehensive overview of the relevant information on the priority pathogens. This included both published information, and the unpublished experience of seed companies and institutes. The surveys also provide an overview of the issues and areas that require further testing or development, and as such are used to aid prioritization of ISHI actions. The main value of the surveys has been to bridge the gaps in knowledge between the participants of the ISHI. The surveys are

Table 12.1. Status report of the International Seed Health Initiative.[a]

Crop	Pathogen	Priority	
		Crop	Pathogen
Apium graveolens (celery)	*Phoma apiicola*	2	2
	Septoria apiicola	2	1
Brassica oleracea	*Phoma lingam*	1	3
	Xanthomonas campestris pv. *campestris*	1	1
Capsicum annuum (pepper)	tobamoviruses	1	1
	Xanthomonas campestris pv. *vesicatoria*	1	1
Cucumis melo (melon)	*Fusarium oxysporum* f. sp. *melonis*	1	3
	SMV	1	1
Cucumis sativus	CGMMV	2	1
(cucumber)	*Pseudomonas syringae* pv. *lachrymans*	2	3
Daucus carota (carrot)	*Alternaria dauci*	2	1
	Alternaria radicina	2	1
	Xanthomonas campestris pv. *carotae*	2	2
Lactuca sativa (lettuce)	LMV	2	1
Lycopersicon esculentum	*Clavibacter michiganensis* pv. *michiganensis*	1	1
(tomato)	*Pseudomonas solanacearum*	1	3
	Pseudomonas syringae pv. *tomato*	1	2
	PSTV	1	3
	tobamoviruses	1	2
	Xanthomonas campestris pv. *vesicatoria*	1	1
Phaseolus spp. (beans)	*Pseudomonas syringae* pv. *phaseolicola*	1	1
	Xanthomonas campestris pv. *phaseoli*	1	1
	Pseudomonas syringae pv. *syringae*	1	2
Pisum sativum (pea)	PSbMV	2	1
	Pseudomonas syringae pv. *pisi*	2	2
	Ascochyta spp.	2	3

[a]Status report is restricted to selected crop species and their associated plant pathogenic microorganisms only and therefore represents a subcomponent of the complete list of priority vegetable species and pathogens.

reviewed regularly as new knowledge becomes available and are also used to develop a set of 'best practices' for preventive seed health management during all stages of the vegetable seed production chain.

Setting up a System for Comparative Testing

The abundance of test protocols for certain pathogens necessitates their systematic comparative testing and evaluation, in order to make a choice of one (or two) recommended 'reference test protocols'. ISHI has developed a procedure for organizing and executing comparative tests, based upon the existing ISTA approach, but described in more detail. Presently there are two types of comparative test: development of a reference method and verification

Table 12.2. Crop–pathogen combinations for which the ISHI is currently preparing a comparative test.

Crop	Pathogen	Date of comparative test
Tomato	*Clavibacter michiganensis* pv. *michiganensis*	1997
	tobamovirus	1997
	Xanthomonas campestris pv. *vesicatoria*	1996–97
Pepper	*Xanthomonas campestris* pv. *vesicatoria*	1996–97
	tobamovirus	1997
Beans	*Xanthomonas campestris* pv. *phaseoli*	1996
	Pseudomonas syringae pv. *phaseolicola*	1996
Watermelon	*Pseudomonas pseudoalcaligenes* subsp. *citrulli* (watermelon fruit blotch)	1997–98

of the suitability of one or two reference methods. In many cases comparative testing will include both types of test, with a total lead time of about 2 years per pathogen. ISHI will perform three to five comparative tests per year.

Execution of Comparative Tests

Two comparative tests have been performed since the start of the ISHI. They evaluated detection of *Xanthomonas campestris* pv. *campestris* in *Brassica* and lettuce mosaic virus in lettuce seed. The technical results from one of these tests are presented in Chapter 27. Table 12.2 shows the crop–pathogen combinations that have a comparative test currently in preparation.

Agreement on a Reference Method

ISHI comparative tests will result in agreement on a 'reference method': an internationally accepted and approved method for the detection of a specific seedborne pathogen. Reference methods will be used as a reference in comparing new methods or in evaluating laboratory performance for detection of a seedborne pathogen. Where possible the method described in the ISTA Working Sheet will be used as the reference method. Agreement on a reference method does not mean that all participants are required to use that method for routine testing. Modified methods or methods developed in-house can be used, although it is advised that these methods are checked for 'equivalency' with the reference method via a (national) laboratory accreditation institute. Official acceptance and recognition of this approach by ISTA is a main objective and will be worked towards during this symposium. It is therefore essential that ISTA and ISHI discuss and agree these procedures.

Agreement on Minimum Threshold

Currently there is no uniform acceptance of what defines a healthy seed. This is considered to be a major stumbling block in international seed trade and in communication with growers. ISHI aims to come to an agreement on a minimum infection threshold for each crop–pathogen combination, which should be adequate for most climatic regions and which can serve as a minimum trade requirement. Additionally it might be necessary to define a more strict threshold for identified 'high risk' climatic regions. Development of minimum thresholds will be one of the major subjects on the ISHI agenda in 1996–97. Owing to a lack of published, sound epidemiological data, these will be mainly based upon the wide practical experience of participating seed companies, inspection services and institutes.

Communication of Results

ISHI aims to develop a clear and effective policy for communication of agreed test protocols, thresholds and other preventive seed health management practices to customers around the world. This will give growers a higher level of confidence in the efforts taken to produce and distribute healthy vegetable seed. The results will also be available to regulatory authorities for the purpose of setting or redefining trade conditions and standards. It is hoped for and estimated that, in the long term, this will lead to more uniformity in the rules and regulations that govern the movement of seeds between countries.

Maintenance and Improvement of Methods

Biotechnological research is continuously resulting in new techniques. These developments will lead to changes in detection methods, in requirements for equipment and materials used by laboratories and in the technical skills required by laboratory personnel. This means that reference methods developed now may soon be outdated and establishing new or improved reference methods will require a new round of comparative testing. It will be a great challenge for both ISTA and ISHI to keep up with these developments.

Quality Assurance

Development of and agreement on a reference method is one achievement, but implementation and execution leading to comparable results between laboratories is a separate issue. ISHI is planning to organize workshops for laboratory personnel of the participating seed companies and institutes/inspection services (early 1998) to discuss these issues. Comparison of laboratory performance is not yet foreseen under the umbrella of ISHI. For the time being this is considered the responsibility of individual laboratories, together with national inspection and/or accreditation bodies. Of course it is essential that performance evaluations are done in a comparable way between countries and ISHI will promote this.

Research into Method and Threshold Development

The comparative tests of 1995/96 have shown that there are still issues where additional research is useful including test development and pathogen epidemiology. When new techniques become available, considerable research will be required before the new method can be used for routine seed health testing. Development of new extremely sensitive techniques will raise new questions regarding acceptable, safe and economically achievable threshold levels, questions that cannot be answered without extensive epidemiological field data.

It is clear that individual seed companies are not able to afford this type of research by themselves. Since the competitive added value of this type of expensive research is limited, there is room for cooperation between seed companies. So far this cooperation has been organized mainly nationally, e.g. in the development of an assay for watermelon fruit blotch (USA), development of a PCR-based assay for bacteria in bean seeds (France) and research into the effect of seed treatment in relation to threshold setting of *Phoma lingam* in *Brassica* seeds (The Netherlands). Further expansion of this cooperation under the ISHI framework is planned.

COOPERATION BETWEEN ISHI AND ISTA

It is clear that cooperation between ISHI and ISTA is a necessity with both organizations profiting from each other's expertise and networks and could lead to an increase in the speed, quantity and quality of output. In this way the interest of worldwide vegetable growers would be best served. Possible areas of cooperation include: (i) comparative testing (standardization of methodology, agreement on priorities and planning, mutual participation, validation of test plans and mutual discussion of results and conclusions); (ii) Working Sheets (evaluation of results and conclusions, publication and adaptation of existing or publication of new Working Sheets); (iii) workshops; and (iv) thresholds (methodology of threshold setting, epidemiological surveys or research and setting and publication of threshold levels).

Cereal Seed Health Strategies in the UK

13

N.D. Paveley[1], W.J. Rennie[2], J.C. Reeves[3], M.W. Wray[3], D.D. Slawson[4], W.S. Clark[5], V. Cockerell[2] and A.G. Mitchell[3]

[1]*ADAS High Mowthorpe, Duggleby, Malton, North Yorkshire YO17 8BP;* [2]*Scottish Agricultural Science Agency, East Craigs, Edinburgh EH12 8NJ;* [3]*National Institute of Agricultural Botany, Huntingdon Road, Cambridge CB3 0LE;* [4]*Pesticides Safety Directorate, Mallard House, Kings Pool, 3 Peasholme Green, York YO1 2PX;* [5]*ADAS Boxworth, Boxworth, Cambridge CB3 8NN, UK*

INTRODUCTION

Cereal production is dependent on a reliable supply of healthy seed. The suppression of seedborne diseases, from epidemic levels in the early part of this century to the point where damage to a commercial crop is a noteworthy event, has been achieved through certification and the use of fungicide seed treatments. It would be wrong, however, to think that absence of visible symptoms implies eradication. Recent survey work (Cockerell and Rennie, 1995) has shown that diseases such as bunt and loose smut are present at low levels in a high proportion of seed. Thus, seedborne diseases have been suppressed to low levels, but the costs are substantial with cereal growers in the UK paying an estimated £23 million per annum for fungicide seed treatments.

Current practice in the use of seed treatments and the procedures of seed certification evolved during a period when organomercury seed treatments were available at negligible cost, and routine use of pesticides was seldom questioned. Concerns about the use of mercury-based pesticides led to the withdrawal of organomercury seed treatments in the UK in 1992. Stimulated by this substantial opportunity, the agrochemical industry introduced a range of novel seed treatments and, as a result, approximately 95% of cereal crops are still grown from treated seed. This provides a baseline from which to re-evaluate seed treatment strategies.

There are several factors to be considered in any re-evaluation:

 Seed Health Testing (eds J.D. Hutchins and J.C. Reeves)

1. A competitive world cereal market continues to demand lower production costs.
2. Political pressure for restraint on pesticide use will focus the industry's attention on implementation of strategies for treatment according to need.
3. The possibility of extending current research in diagnostic techniques to the development of highly sensitive seed health tests, which with appropriate sampling techniques might ultimately provide instant results 'off the back of the combine'. Rapid laboratory-based tests are a realistic prospect in the short to medium term.

If such tests could provide accurate information on seed health status and these data were quickly and readily available, growers might reasonably question whether the treatment of seed known to be free of disease aided the suppression of seedborne diseases. If not, is the significant cost of treatment justified by other potential benefits? There is a risk that new testing techniques will be developed before the industry understands how to convert the results into appropriate decisions. Hence the seed health debate needs to guide the technology through the consideration of future seed treatment strategies. The main aims of any strategy to maintain cereal seed health are to: (i) ensure long-term security of production against the deleterious effects of seedborne diseases on yield and grain quality; (ii) minimize input costs; (iii) maximize any additional benefits of seed treatment against soilborne or foliar diseases; (iv) ensure adequate returns to agrochemical companies and plant breeders to stimulate development of better seed treatments and varieties; and (v) minimize risks to operators, consumers and the environment. The first of these aims cannot be compromised, and the high multiplication potential of seedborne diseases does not allow the current level of suppression to be relaxed. However, it is reasonable to ask if seed treatments can be targeted more effectively, while achieving a balance between the other needs. The cereal industry has a collective responsibility for seed health, and must consider how to maintain this as economic, political and technical factors change.

The review from which this paper was derived (Paveley *et al.*, 1996) brings together information on seed treatment and seed production practice in the UK, and relates it to our understanding of disease epidemiology. It asks whether current practice is the most effective way to manage seedborne diseases in the post-organomercury era.

THE MAIN SEEDBORNE DISEASES OF CEREALS IN THE UK

A survey carried out by Cockerell and Rennie (1995) from 1992 to 1994 identified bunt and *Fusarium* seedling blight in wheat, and loose smut and leaf stripe in barley as the most damaging or potentially damaging seedborne diseases of cereals in the UK.

Bunt

Cockerell and Rennie (1995) tested over 200 samples of winter wheat seed in each of the years from 1992 to 1994. Spores of *Tilletia caries* were recorded in 20–60% of samples, suggesting widespread contamination of UK winter wheat seed. However, the level of contamination was low, with only 2–8% of samples having more than one spore per seed.

Fusarium Seedling Blight in Wheat

Fusarium seedling blight, caused mainly by *Fusarium culmorum* and *Microdochium nivale*, is common in the UK. Severe infection may result in poor plant establishment. Cockerell and Rennie (1995) found that 99% of winter wheat seed samples carried some *M. nivale* in 1992 and 1993 compared with 68% of samples in 1994. This probably reflects the occurrence of weather conditions conducive to infection of ears. The severity of infection also varied greatly during the three years of the survey. In 1992 and 1993, over 40% of certified seed samples had at least 20% of seeds infected, but in 1994 only 2% of samples exceeded this level of infection.

Loose Smut of Barley

Cockerell and Rennie (1995) found that farm-saved seed was more frequently infected with *Ustilago nuda* and at higher levels than certified seed. Winter barley was more often and more heavily infected than spring barley, reflecting the greater susceptibility of winter varieties. In 1992 and 1993, 1% of samples of winter barley certified seed had more than 0.5% infection (the European Union standard for certified seed), but in 1992, 25% of farm-saved samples had this level of infection.

Leaf Stripe

During the mid-1980s there were occasional reports of leaf stripe in spring barley crops grown from organomercury treated seed in the UK. Jones *et al.* (1989) later confirmed that some isolates of *Pyrenophora graminea* were resistant to organomercury. In the period from 1987 to 1992, the highest incidence of *P. graminea* in Scotland was recorded in spring barley seed harvested in 1989 and 1990 (Cockerell *et al.*, 1995). Winter barley had a much lower incidence and severity of infection. In 1989 and 1990, 69% and 82% of samples of spring barley seed respectively were infected, with a mean of 6% of seeds infected. Between 1990 and 1992, infection decreased following the introduction of a voluntary Code of Practice for the sale of infected seed (Anon., 1991), and a move away from organomercury fungicides to alternative seed treatments.

CURRENT PRACTICE FOR TREATMENT AND DISEASE TESTING OF CERTIFIED SEED

Fungicides are routinely applied to certified cereal seed in the UK, although effective treatment is not a requirement of certification. The seed industry views routine treatment as a good insurance against the effects of seedborne diseases and potential claims for compensation. In addition, the income from seed treatments is important to seed companies. Very little disease testing is done on certified seed, although some companies routinely test seed of susceptible barley varieties for loose smut before applying seed treatments.

CURRENT PRACTICE FOR FARM-SAVED SEED

The use of farm-saved seed in the UK is about 20–30%, estimated from the national cereal area, assumed seed rates and the total weight of certified seed. Survey data (Polley *et al.*, 1995a,b) agree with this estimate. Quality assurance measures for farm-saved seed vary greatly, from nothing at all to procedures similar to (and in some respects even more rigorous than) those for certified seed. In the majority of cases certified seed, often first-generation (C1), is used to sow the crop used for farm-saved seed. Very few seed tests are done, except for germination and moisture content. Most of the untreated seed sown in the UK is farm saved (Table 13.1).

REASONS FOR SEED TREATMENT

Control of Seedborne Diseases

The primary reason for fungicide treatment of cereal seed is to control seedborne diseases. However, decisions on the use of seed treatments are complex, and few cereal producers have a good understanding of the reasons for applying them. A large proportion take little interest in which treatment is applied to their seed, and leave the decision to the seed merchant. This was understandable when treatment with organomercury cost £6–8 t^{-1}, but the

Table 13.1. Treatment of certified and farm-saved seed, Cereal Disease Survey 1995 (Polley *et al.*, 1995a,b)

	Certified seed		Farm-saved seed		
	Treated (%)	Untreated (%)	Treated (%)	Untreated (%)	No. of crops
Winter wheat	99	1	92	8	360
Winter barley	98	2	81	19	351

wide range of alternative treatments cost £38–85 t^{-1}, and it is surprising that many farmers continue to take little interest in seed treatment decisions.

Control of Soilborne Diseases

Bunt

Soilborne bunt has until recently been of little concern in the UK. Free bunt spores in soil germinate in the presence of moisture and if no host crop is present they die. Thus, under normal UK autumn conditions, spores shed to the soil of the harvested or adjacent fields would germinate before a following wheat crop was sown into that field. Occasional bunt outbreaks due to soilborne infection have been confirmed when soil conditions between harvest and sowing were very dry (Yarham, 1993). Those farmers, and their neighbours, who have had soilborne bunt or severe infection leading to soil contamination, must rely on seed treatments which are active against soilborne infection.

Fusarium seedling blight

Experimental work by Bateman (1977), using artificial inoculation, suggested that soilborne *Fusarium* was a possible source of seedling blight. Before work carried out by ADAS (described below) the relative importance of natural soilborne and seedborne inoculum had not been quantified. As a result of this uncertainty, even when seed had been tested and found to be within acceptable tolerances, the risk of soilborne infection made it 'advisable to treat where there is a risk of winter crops being sown late, or where seed may be sown in unusually cold or wet seedbeds that may delay seedling emergence' (Rennie and Cockerell, 1993). Since seedbed conditions cannot be predicted at the time of the treatment decision, most cereal growers err on the side of caution and use treated seed.

Control of Foliar Diseases

Control of foliar diseases by application of a seed treatment has often been the goal of agrochemical manufacturers. When first introduced, Milstem (ethirimol) gave excellent control of barley powdery mildew. Subsequently, Ferrax (ethirimol + flutriafol) gave good foliar disease control, and is still used on some barley crops where early mildew control is required. In wheat, the use of seed treatments for the control of foliar diseases is almost totally confined to the control of yellow rust.

Seed Treatment in the Absence of Seedborne Diseases

There is a clear distinction between the use of seed treatments to control seedborne diseases and their use on seed that is known to be healthy. A series of experiments was carried out by ADAS to see if treatment of seed shown to be within acceptable tolerances for seedborne pathogens is justified by the potential benefits from control of soilborne and foliar diseases, and other effects such as reduced lodging. The experiments at 23 sites over 3 years used seed stocks shown to be substantially free of seedborne diseases. There was considerable site and seasonal variation in plant establishment, ranging from approximately 100 to 400 plants m^{-2}. Regression analysis showed that untreated establishment explained 87% of the variation in treated establishment and there was a small overall benefit from treatment. However, this small benefit could be explained by control of a low level of seedborne *Fusarium* seedling blight on one batch of seed. No significant control of foliar disease was detected by disease assessments at growth stage 31 (Tottman, 1987).

These results support earlier work by Richardson (1986) suggesting that the risk to crop establishment from natural soilborne *Fusarium* is negligible. In the absence of seedborne contamination, the effect of treatment is as likely to be deleterious as positive, even in poor seedbed conditions. Attention can therefore be focused on the ability of seed tests to differentiate between infected and healthy seed.

SEED TESTING

The preference for growing winter cereals in England and Wales creates a large demand for seed in late September and early October. Over 80% of winter wheat crops are drilled between 21 September and 31 October, and over 80% of winter barley between 11 September and 20 October (Polley *et al.*, 1995a,b). This large autumn peak demands quality assurance systems that can cope with a very high throughput, without delaying the supply of seed. Any strategy for treatment based on seed health tests must take this into account.

Seed health tests have a number of general requirements which were reviewed by Reeves (1995). These include speed, particularly where high throughput is needed, and simple procedures with as few steps as possible to minimize handling and errors. Tests should be accurate and repeatable within limits of variation, easy to record objectively, and not too costly. Establishment of the relationship between laboratory test results and the incidence of disease in the growing crop is crucial.

Current Seed Testing Methods

Current seed testing technology for fungal pathogens has been used for many years, and although it can be extremely effective and appropriate in some contexts, it has a number of disadvantages. Firstly, incubation tests are slow, typically taking a week or more. Secondly, the identification of a particular pathogen requires skill and experience. Some pathogens are especially difficult to identify, e.g. *Pyrenophora* spp., and their presence may be obscured by other fungi and bacteria growing on the seed.

The disadvantages of existing methods mean that seed health tests are relatively expensive (appoximately £40–50 per test at the Official Seed Testing

Station, Cambridge, 1997). Large-scale testing with high throughput requires significant investment to provide sufficient capacity and trained personnel. This has led to much interest in the use of new techniques to improve the speed, accuracy and sensitivity of diagnosis (Reeves, 1995).

New Test Methods

Immunodetection

Recent developments in seed health testing have concentrated on immunodetection techniques. Early immunoassays used polyclonal antibodies, but monoclonal antisera are increasingly being used to improve accuracy and overcome unwanted cross-reactions. Further developments will probably allow these assays to be used in the field as 'dipstick' type tests, as well as in the laboratory. Antibody-based tests have been developed for detecting cereal pathogens (Dewey and Priestley, 1994), and some of these have targeted seedborne pathogens with varying degrees of success (Burns *et al.*, 1994). None is yet being used as part of routine high throughput seed health testing.

Polymerase chain reaction

The most powerful new technology available for the improvement of seed health testing is based on the molecular biochemistry of nucleic acids. These techniques have progressed rapidly over the last decade, and have a great power of discrimination between organisms and can detect them at very low concentrations. Such attributes are valuable for seed health testing and have the potential to overcome some of the disadvantages of conventional methods.

Although DNA probes are highly specific to a particular pathogen, problems have been encountered with their sensitivity and the presence of contaminating DNA. In the mid-1980s, the polymerase chain reaction (PCR) was developed (Saiki *et al.*, 1985). This technique has the potential to combine the specificity of DNA probes with a high level of sensitivity, making it the preferred method for new tests to detect pathogens in medical, food, plant and, increasingly, seed pathology. Consequently, the use of PCR for diagnostic applications in plant and seed pathology is growing rapidly (Schots *et al.*, 1994; Reeves, 1995).

There are, however, some disadvantages to PCR. Whilst it is reliable when amplifying from pure DNA, the reaction may be inhibited by other materials. Problems have been encountered with some seed material (Reeves, 1995), although PCR has been made to work in the presence of biological matrices such as milk, saliva, cerebrospinal fluid and sewage sludge. The problem of inhibition is not insuperable but adds to development costs.

INTERPRETATION OF SEED TEST RESULTS AND THRESHOLD DEFINITION

The value of seed health testing lies in its ability to support decisions about suitability for seed and the need for treatment. Two test result thresholds may be set, the lower below which seed can be sown untreated, and the higher above which treatment cannot provide robust control at an economic cost. The information required to set thresholds depends primarily on the epidemiology of the pathogen and how the seed will be used. The aim should be to set thresholds at levels which, on average, maximize profit, minimize the risk of occasional severe loss in an individual crop and place the pathogen population under constant downward pressure.

For most seedborne diseases, the relationship between test result and disease expression, the risk of spread to neighbouring crops and the rate of generation on generation increase in seed infection all interact with agronomic and environmental conditions. For example, the expression of seedborne *Fusarium* seedling blight is influenced by drilling date and seedbed conditions. Hence, for a given test result, the interactions lead to a range of possible outcomes, which must be properly quantified if thresholds are to be robust but not over-cautious. In practice, many of the thresholds currently used to convert test results into decisions have been developed empirically without key experimental data.

RATIONALIZING SEED TREATMENT USE

No pesticide use is entirely free of risk. The most obvious sources of public and user concern are operator exposure during seed handling and drilling, poisoning of farm and domestic livestock from accidental consumption of treated seed, and intake of treated seed by wild birds. All require consideration in any debate on future strategies for seed treatment use. The notion that pesticides should be used on a rational 'according to need' basis has been accepted by the agricultural industry. In the case of foliar sprays, the substantial cost benefits of moving from routine prophylactic treatment to treatment based, for example, on varietal resistance and disease risk, have encouraged the uptake of new techniques. The potential economic benefits from rationalizing seed treatment use are smaller, but savings are possible on the £23 million per annum currently spent on cereal seed treatments in the UK.

POSSIBLE FUTURE STRATEGIES FOR UK SEED HEALTH AND SEED TREATMENT USE

Four possible strategies for the future maintenance of seed health are outlined in the following sections, together with a brief summary of their implications. It should be emphasized that these strategies are proposed for discussion and do not represent the policy of government departments or the cereal industry.

Minimal Change to the Current Practice of Routine Seed Treatment

A case can be made on technical and commercial grounds to continue the current practice of routine use and to maintain the certification standards for loose smut and ergot. There are several advantages to this case:

1. It can be argued that current practice has achieved and maintained a high level of seed health and ensured minimal losses from seedborne diseases. Although a small increase in the amount of untreated seed would probably make little difference to this position, significant changes may have unforeseen consequences on the incidence and severity of seedborne diseases.
2. The short time available for cleaning, testing and treating seed of winter cereals would make it difficult to change current practice given the constraints of existing seed testing technology.
3. The profit generated from sale of fungicide seed treatments for seed merchants, agrochemical companies and mobile seed processors is of sufficient importance for there to be resistance to change.

On the other hand there are several disadvantages:

1. Growers will continue to bear the costs of routine seed treatment use and, indirectly, certification controls for loose smut and ergot.
2. A substantial proportion of cereal seed with nil or sub-threshold levels of seedborne disease will continue to be treated unnecessarily.
3. Growers may respond to falling margins by farm-saving a higher proportion of their seed and leaving it untreated, leading to a gradual increase in the use of untreated seed without appropriate tests.

Limited 'Treatment According to Need' Strategy

A change to a limited 'treatment according to need' approach to the use of fungicide seed treatment in spring cereals, both certified and farm-saved seed, is feasible with existing seed testing techniques, and could be implemented in the short term. This strategy might be particularly appropriate in Scotland, where spring barley occupies approximately 60% of the cereal area.

Long-Term Change Towards a Wider Strategy of 'Treatment According to Need'

Better targeting of seed treatments, both in avoiding treatment of healthy seed and in applying the products most appropriate to the range of pathogens present, could be achieved through advances in seed testing technology and an increase in understanding of the major seedborne diseases. A significant long-term decrease in the use of fungicide seed treatments in cereals should be achievable, while maintaining or improving seed health. Such a strategy could begin to take practical effect in approximately 5 years given the following advances: rapid, sensitive and reliable tests for seedborne diseases would be necessary to meet the demand for processing of winter cereals in September and October. DNA-based tests could meet these requirements, and their development is feasible with current knowledge. Automation in testing procedures could reduce unit costs and allow testing organizations to cope with a high throughput of samples. Quantitative information would be needed to allow the results of seed health tests to be accurately related to the risk of seedborne disease development in the field. Seed testing in the certification process could replace visual inspections for disease in control plots, and broaden the range of diseases for which thresholds for treatment or rejection could be set. However, the main disadvantage concerns the reduced seed treatment market which may prejudice the commercial viability of developing new fungicides; the cost of certified seed may increase to compensate for the loss of income from seed treatments.

Treatment Based on Compulsory Testing of all Seed

Given the 'social' aspects of seedborne diseases, it might be argued that long-term compliance with rational control strategies can only be ensured by legislation. Such legislation might make treatment of certified and possibly farm-saved seed dependent on the results of seed health tests. It is unlikely that the legislation required to implement this strategy would obtain political support or be commercially acceptable, given the current emphasis on deregulation. However, some elements could be cost-effectively introduced into the certification process.

CONCLUSIONS

The certification process and routine use of cereal seed treatments in the UK has ensured a supply of healthy seed, but there are strong reasons for re-evaluating this strategy. A limited strategy of 'treatment according to need' is possible with existing seed testing technology. However, a significant shift to treatment based on the results of seed health testing will require new seed health tests, knowledge of how to use them, and a better understanding of key aspects of the epidemiology of the most important diseases.

ACKNOWLEDGEMENTS

The authors acknowledge the support of the Home-Grown Cereals Authority in funding the review from which this paper was abstracted. Thanks are due to colleagues in ADAS, the National Institute of Agricultural Botany, the Scottish Agricultural Science Agency, and the Central Science Laboratory for the provision of data, information and advice.

REFERENCES

Anonymous (1991) *Barley Leaf Stripe – Pyrenophora graminea – A Code of Practice for Seed Intended for Sale.* The National Farmers' Union for Scotland, The Scottish Council of UKASTA and the Scottish Seed Growers' Association.

Bateman, G.L. (1977) Effect of organomercury seed treatment of wheat seed on *Fusarium* seedling disease in inoculated soil. *Annals of Applied Biology* 85, 195–201.

Burns, R., Vernon, M.L. and George, E.L. (1994) Monoclonal antibodies for the detection of *Pyrenophora graminea*. In: Schots, A., Dewey, F.M. and Oliver, M. (eds), *Modern Assays for Plant Pathogenic Fungi.* CAB International, Wallingford, UK pp. 199–203.

Cockerell, V. and Rennie, W.J. (1995) *Survey of Seed-borne Pathogens in Certified and Farm-saved Seed in Britain between 1992 and 1994.* Home-Grown Cereals Authority Project Report no. 124, Home-Grown Cereals Authority, London.

Cockerell, V., Rennie, W.J. and Jacks, M. (1995) Incidence and control of barley leaf stripe (*Pyrenophora graminea*) in Scottish barley during the period 1987–1992. *Plant Pathology* 44, 655–661.

Dewey, F.M. and Priestley, R.A. (1994) A monoclonal antibody-based technique for the detection of the eyespot pathogen of cereals *Pseudocercosporella herpotrichoides.* In: Schots, A., Dewey, F.M. and Oliver, M. (eds). *Modern Assays for Plant Pathogenic Fungi.* CAB International, Wallingford, UK, pp. 9–15.

Jones, D.R., Slade, M.D. and Briks, K.A. (1989) Resistance to organomercury in *Pyrenophora graminea. Plant Pathology* 38, 509–513.

Paveley, N.D., Rennie, W.J., Reeves, J.C., Wray, M.W., Slawson, D.D., Clark, W.S., Cockerell, V. and Mitchell, A.G. (1996) *Cereal Seed Health Strategies in the UK.* Home-Grown Cereals Authority Research Review no. 34, Home-Grown Cereals Authority, London, 131 pp.

Polley, R.W., Slough, J.E. and Jones, D.R. (1995a) Survey of winter wheat diseases in England and Wales – 1995. Internal Report, CSL Harpenden and ADAS.

Polley, R.W., Slough, J.E. and Jones, D.R. (1995b) Survey of winter barley diseases in England and Wales – 1995. Internal Report, CSL Harpenden and ADAS.

Reeves, J.C. (1995) Nucleic acid techniques in testing for seed borne diseases. In: Skerrit, J.H. and Appels, R. (eds), *New Diagnostics in Crop Sciences.* CAB International, Wallingford, pp. 127–149.

Rennie, W.J. and Cockerell, V. (1993) A review of seed-borne pathogens of cereals in the post-mercury period. *Proceedings Crop Protection in Northern Britain*, pp. 17–24.

Richardson, M.J. (1986) An assessment of the need for routine use of organomercurial cereal seed treatment fungicides. *Field Crops Research* 13, 3–24.

Saiki, R.F., Scharf, S., Faloona, F., Mullis, K.B., Horn, G.T., Erlich, H.A. and Arnheim, N. (1985) Enzymatic amplification of β-globin genomic sequences and restriction site analysis for diagnosis of sickle cell anaemia. *Science* 230, 1350–1354.

Schots, A., Dewey, F.M. and Oliver, M. (1994) *Modern Assays for Plant Pathogenic Fungi*. CAB International, Wallingford, UK, 267 pp.

Tottman, D.R. (1987) The decimal code for the growth stages of cereals, with illustrations. *Annals of Applied Biology* 110, 441–454.

Yarham, D.J. (1993) Soil-borne spores as a source of inoculum for wheat bunt (*Tilletia caries*). *Plant Pathology* 42, 654–6.

Review on Policy Developments with Regard to Seed Health Testing and Seed Treatment in the Nordic Countries with Special Reference to Denmark

14

C. Scheel

Danish Plant Directorate, Skovbrynet 20, DK-2800 Lyngby, Denmark

INTRODUCTION

Many countries have a general policy aiming to reduce the use of pesticides, often driven by environmental reasons. Fungicide treatment of seed is an area where it may be possible to reduce the quantity of chemicals used by limiting treatment according to need. By relatively simple tests for seedborne diseases, treatment may be limited to infected lots only. Cereal seeds are produced in relatively large quantities and, therefore, if only a small proportion of seed lots require treatment, then the quantities of fungicides used could be reduced.

In Sweden, regulations limiting the use of seed treatments for cereals were introduced in 1965 (Brodal *et al.*, 1994). Today seed health testing of cereal seed is compulsory in Sweden, but the use of fungicide treatments is based on recommendations. According to Brodal (1996) a voluntary scheme for the reduction of the use of seed treatments was introduced in Norway in 1990, and led to a significant reduction in the use of such products. In contrast, Finland has not taken similar initiatives (Brodal *et al.*, 1994). Since 1971, the Danish Plant Directorate (formerly the Danish State Seed Testing Station) has offered to carry out tests to assess the need for seed treatment of Danish spring cereals (Anon., 1971). However, only a few seed producers have taken advantage of this offer which consequently has not led to any significant reduction in the use of seed treatments. At present about 90% of all certified cereal seed in Denmark is routinely treated with fungicides.

(eds J.D. Hutchins and J.C. Reeves)

In 1995 the Danish Ministry of Environment and Energy in collaboration with the Ministry of Agriculture and Fisheries formed a working group to assess the possibility of reducing the use of fungicide-treated cereal seed in Denmark. This group consisted of representatives from The Danish Plant Directorate, the Danish Research Service for Plant and Soil Science, seed companies and farmers' organizations. The basis for the discussion was that all seed lots of the first three seed generations (pre-basic, basic and certified first generation seed) should be treated with fungicides. Only seed of the certified second generation, constituting about 80% of the total quantity of seed produced in the certification scheme, should be treated according to need.

SEEDBORNE DISEASES IN CEREALS

The important cereal seedborne fungal pathogens in Denmark are listed in Table 14.1. Expected yield losses without the control of these pathogens are given as well as estimates of the effect of seed treatment on seedborne inoculum.

Monocyclic Seedborne Diseases

The following monocyclic seedborne diseases are of potential importance under Danish conditions: *Drechslera graminea* (barley leaf stripe), *Tilletia tritici* (wheat bunt), *Urocystis occulta* (rye stem smut), *Ustilago* spp. (loose and covered smuts of barley and oats) and *Ustilago tritici* (loose smut of wheat and barley). The latter is not controlled by the fungicides in general use in Denmark and is therefore not included in the present discussion.

If all cereal seed of the first three generations (pre-basic, basic and certified first generation) are treated effectively with fungicides against these seedborne diseases, then it may be assumed that very little infection will be found in the final certified second generation seed. However, there are some situations where the risk of infection must be taken into consideration in deciding if seed of the certified second generation should be tested for disease, before being sown without a seed treatment:

1. Spores from neighbouring barley fields infected by *D. graminea* (leaf stripe), may be spread by wind to fields containing a certified first generation crop and thereby contaminate seed to be used for the certified second generation. However, leaf stripe occurs very sporadically at present in Denmark and the risk of contamination from infected fields is small. But if the disease occurs, severe attacks can develop and a test for leaf stripe in the certified second generation barley seed would be recommended.
2. Spores of *T. tritici* (wheat bunt) and *U. occulta* (rye stem smut) may be soilborne and airborne (spread during threshing and processing). Likewise covered smuts of *Ustilago* spp. may be spread during threshing and processing.

Table 14.1. Important seedborne fungal pathogens in cereals in Denmark.

Cereal species	Disease	Expected yield loss at heavy attacks[a]	Effect of seed treatment[c]
Wheat	bunt (*Tilletia tritici*)	100%[b]	***
	glume blotch (*Septoria nodorum*)	5–10%	**
	seedling blight (*Fusarium* spp.)	5%	**
Barley	leaf stripe (*Drechslera graminea*)	1% per percentage of infected plants	***
	net blotch (*Drechslera teres*)	Insignificant	***
	seedling blight, spot blotch (*Cochliobolus sativus*)	3%	**
	seedling blight (*Fusarium* spp.)	3%	**
	loose smut (*Ustilago tritici*)	0.75% per percentage of infected plants	**
Oat	seedling blight (*Fusarium* spp.)	5%	**
Rye	stem smut (*Urocystis occulta*)	1% per percentage of infected plants	***
	seedling blight (*Fusarium* spp.)	5%	**

[a]Expected yield losses at heavy attacks can vary considerably from year to year.
[b]100% loss in quality because the cereal is of no value due to the bad smell.
[c]**, moderate effect of fungicide seed treatment on seedborne inoculum; ***, maximal effect of fungicide seed treatment on seedborne inoculum.

Bunt has appeared lately in several Danish wheat fields, mainly because some of the fungicides used to replace mercury dressings are not sufficiently effective. Therefore at present, wheat seed of the certified second generation should not be sown untreated unless a laboratory test has proven the seed to be free of the pathogen.

The other diseases listed above are at present of little importance in Denmark and seed of the certified second generation can be sown untreated without any risk, if the first three generations were treated effectively.

Polycyclic Seedborne Diseases

The most important polycyclic seedborne diseases on cereals in Denmark are *Cochliobolus sativus* (seedling blight, spot blotch), *Drechslera teres* (net blotch of barley), *Fusarium* spp. (seedling blight) and *Septoria nodorum* (glume blotch).

Seed treatment may prevent these fungi from affecting seed germination, but may have little effect on disease development later in the growing season, as inoculum of other sources will often be present in the field. In some years (under Danish conditions) winter wheat seed may be heavily contaminated with *S. nodorum* and *Fusarium* spp. These pathogens may seriously harm germination and untreated seed should not be sown, unless a test has proven that the level of infection is low. In very wet years, which occur infrequently, barley seed may be heavily contaminated by *C. sativus* and *Fusarium* spp. In most years, however, seed could be sown untreated without risk of seedling attack by these fungi.

The levels of *D. teres* on barley seed vary from year to year. The fungus has little effect on seedling development. However, if the seed is sown in soil free from inoculum of the fungus, seed treatment may delay development of a net blotch epidemic in the field. The seed health testing method presently applied does not distinguish between infection of *D. teres* and *D. graminea* (see Chapter 2).

PRACTICAL CONDITIONS IN RELATION TO SEED HEALTH TESTING

Sampling

A sample has to be representative of the entire lot from which it is removed if testing to assess the need for fungicide treatment is the basis for a more rational use of seed treatments. Sampling a seed lot is normally presumed to give a realistic impression of the frequency of disease in the entire lot.

If the harvested seed is stored on the farm, samples must be drawn by hand during the movement of the seed, because the storage bins are usually too deep and wide for hand sampling once in storage. Automatic sampling equipment that could function under a variety of conditions would be useful in this situation. Another possibility could be direct sampling on the combine; however, the technology of the sampling equipment required needs further development. If samples are taken after the seed has been delivered to the seed merchant, then the merchant will have to provide facilities to store the seed lots in smaller single units to allow sampling of each lot, rather than bulking seed lots together. In Denmark, most seed merchants do not have the facilities to store large quantities of seed in this manner. It would require enlargement of their storage infrastructure, which is unfeasible for the majority of merchants.

Testing

The methods used at the Danish Plant Directorate to test seed to assess the requirement for fungicide treatment are listed in Table 14.2, together with the duration of the laboratory tests. Most of the methods have been developed and recommended by the International Seed Testing Association (ISTA). The test methods and principles are as follows:

- Doyer's method: after incubation the percentage of seedlings with discoloured roots is counted.
- Fluorescent method: after incubation the percentage of kernels with fluorescent mycelium is counted.

Table 14.2. Methods used at the Danish Plant Directorate to test cereals for seedborne pathogens.

Cereal species	Pathogen	Testing method	Duration in laboratory
Spring cereals			
Spring wheat[a]	*Fusarium* spp.	Doyer's method	14 days
	Septoria nodorum	fluorescent method	10 days
Spring barley[b]	*Cochliobolus sativus*	Doyer's method	7 days
	Fusarium spp.		
	Drechslera graminea/D. teres	freezing blotter or osmotic method	7 days
Oat	*Fusarium* spp.	Doyer's method	7 days
Winter cereals			
Winter wheat	*Fusarium* spp.	Doyer's method	14 days
	Septoria nodorum	fluorescent method	10 days
	Tilletia tritici	washing method	2 hours
Winter barley[b]	*Cochliobolus sativus*	Doyer's method	7 days
	Fusarium spp.		
	Drechslera graminea/D. teres	freezing blotter or osmotic method	7 days
Rye[c]	*Fusarium* spp.	Doyer's method	7 days

[a]Spring wheat is not tested routinely for *Tilletia tritici.*
[b]Barley is not tested routinely for *Ustilago tritici.*
[c]Rye is not tested routinely for *Urocystis occulta.*

- Freezing blotter method: after incubation the percentage of kernels with spores of *Drechslera* sp. is counted.
- Osmotic method: after incubation the percentage of kernels where *Drechslera* sp. has produced a coloured pigment is counted.
- Washing method: spores are washed off the seeds and counted with the aid of a haemocytometer.

The freezing blotter method is used routinely in Denmark when testing for leaf stripe/net blotch (*D. gramine/D. teres*) in barley. In Norway and Sweden it is replaced by the osmotic method, which is less laborious (see Chapter 2).

In general there is a problem of a time constraint when testing winter cereals. High inputs of resources are required in order to complete all of the necessary tests in the short period between harvest and sowing. The testing of winter wheat takes approximately 14 days, which is too long considering the available window of opportunity. In practice it is only possible to test winter wheat seed for bunt; however, this test alone is not sufficient to determine the need for seed treatment.

Table 14.3. Threshold levels for seed treatment set by the Danish Plant Directorate.

Cereal species	Pathogen	Recommended threshold levels for seed treatment
Spring cereals		
Spring wheat	*Fusarium* spp.	15%
	Septoria nodorum	5%
Spring barley	*Drechslera graminea/D. teres*	5%
	Cochliobolus sativus	30%
	Fusarium spp.	
Oat	*Fusarium* spp.	15%
Winter cereals		
Winter wheat	*Fusarium* spp.	15%
	Septoria nodorum	5%
	Tilletia tritici	spores present
Winter barley	*Drechslera graminea/D. teres*	5%
	Cochliobolus sativus	15%
	Fusarium spp.	
Rye	*Fusarium* spp.	15%

Threshold Levels

The threshold levels set by The Danish Plant Directorate for seed treatment in Denmark are listed in Table 14.3. These limits for deciding when seed treatments are recommended are based on research and experience. In Norway there have been seedborne infections in barley seed ranging from 25% to 78% (average 60%) and for wheat seed between 35% and 90% (average 70%), where seed treatment has been recommended in the period from 1990 to 1994 (Brodal, 1995). Experiences from the Danish Plant Directorate have shown an even lower percentage of need for fungicidal treatment of spring cereals (Jørgensen, 1986, 1988, 1990, 1996).

It is expected that the wider practice of reduced seed treatment according to need could lead to a higher disease occurrence, and that it may be necessary to change the threshold levels to lower limits in the future. Diseases that are uncommon or of minor importance today could become more frequent and lead to alternative strategies in relation to testing seed to assess the need for fungicide treatment.

CONCLUSIONS – POSSIBILITIES OF REDUCING THE USE OF FUNGICIDE-TREATED CEREAL SEED

The working group formed by the Danish Ministry of Environment and Energy together with the Ministry of Agriculture and Fisheries concluded the following:

1. Only spring cereals can be tested routinely to assess the need for fungicide treatment. Most winter cereals have to be treated without testing, due to the lack of time in the period between harvest and sowing.
2. All certified seed of pre-basic, basic and certified first generation spring cereals has to be treated with approved fungicides.
3. Certified second generation seed of spring cereals should be tested for the presence of fungal pathogens and the results used to indicate the need for seed treatment.

An essential problem in implementing such a system is the inability to draw representative samples from individual seed lots. The seed is often stored in (bulked) facilities on the farm from which it is not possible to draw representative samples. If the samples are to be taken after the seed is delivered to the seed merchant, then it is necessary for the merchant to change the methods they use for handling large quantities of seed. After cleaning, the seed should be stored in small containers (1.5–2 tonnes); representative samples could then be taken from these smaller lots. Such a change is not possible at the moment. Seed merchants are generally held responsible for attacks of seedborne diseases in the field and therefore tend to treat seed routinely rather than test the seed to assess the need for treatment.

In Denmark, if spring cereal seed is treated only when required it will lead to a reduction in the use of chemical fungicides. Approximately 100,000 tonnes of chemicals were applied to seed of the certified second generation of spring cereals in 1995. This equates to approximately 5 tonnes of fungicide active ingredient. With the need for seed treatment varying from year to year from 25% to 90% of seed lots tested (the Norwegian experience), a reduction in use of chemicals will result in reductions of 0.5–3.8 tonnes of active ingredients per year.

The increased cost required to establish sampling equipment and for sampling and testing each seed lot is about equal to the reduction in expenses for chemical treatment. Without a greater economic stimulus there is currently no basis for the establishment of compulsory seed health testing of cereals in Denmark (Anon., 1995). A positive effect on the environment due to the reduction in use of chemicals is thus the only argument for implementing seed treatment according to need in Denmark.

REFERENCES

Anonymous (1971) Cirkulære til sædekornsfirmaerne vedrørende undersøgelser ved Statsfrøkontrollen af behov for bejdsning af udsæd af byg. *Statsfrøkontrollens Cirkulære*, pp. 216–271.

Anonymous (1995) *Rapport fra Arbejdsgruppen vedrørende undersøgelse af mulighederne for reduktion af brugen af bejdsemidler til sædekorn.* Plantedirektoratet.

Brodal, G. (1995) Behovsprøvet såkornbeising – Miljømessige og økonomiske fordeler etter fem år med systematiske analyser. *Landbrukstilsynets Rapport om Frøkontrollvirksomhet* 1994/95, pp. 65–70.

Brodal, G. (1996) Economic and ecological benefits of seed health testing in Norway. Presentation at the ISTA Pre-Congress Seminar on Seed Health Testing in the Production of Quality Seed, Copenhagen (in press).

Brodal, G., Halkilahti, A.M., Jørgensen, J. and Sperlingsson, K. (1994) *Frøbårne Plantesjukdommer og Frøpatologiske Undersøkelser.* TemaNord 1994. Nordisk Ministerråd, Copenhagen, 630 pp.

Jørgensen, J. (1986) The incidence of *Cochliobolus sativus* on seed samples of barley and oat in Denmark during 20 years related to climatical conditions and other factors. *Report from the Danish State Testing Station* 115, 107–119.

Jørgensen, J. (1988) Om forekomster af *Fusarium* spp. på udsæd af vårbyg i Danmark og om behovet for bejdsning. *Report from the Danish State Testing Station* 117, 147–152.

Jørgensen, J. (1990) Om behovet for bejdsning af sædekorn. *Landbonyt*, March 1990, 44–47.

Jørgensen, J. (1996) *Plant Pathological Records in the Annual Reports of the Danish State Seed Testing Station 1871–1990.* Dyssemarken 10, Måløv, Denmark.

Developments in Seed Health Testing Policy – a US Perspective

15

D. McGee

Seed Science Center, Iowa State University, Ames, IA 50011, USA

INTRODUCTION

Transmission of plant pathogens by seeds has economic impact on the USA seed industry in meeting export phytosanitary certification standards and liability regarding quality of seeds and transmissibility of pathogens to new crops. Two basic methods are used to certify seeds in order to meet the phytosanitary requirements of exported seed lots, namely field inspection or laboratory assays. The US Department of Agriculture (USDA) Animal and Plant Health Inspection Service (APHIS) is responsible for the phytosanitary certification on seeds in the USA and authorizes public agencies in the states to carry out the appropriate testing of seeds. The current international phytosanitary system, however, does not mandate any standards for seed health testing either in the field or laboratory.

Until 1988, seed health issues were perceived quite differently by the vegetable and field crop seed companies. Vegetable companies recognized the need to assay seeds for economically important seedborne pathogens such as *Xanthomonas campestris* on brassicas, lettuce mosaic virus, bacterial blights of beans and many others. Seed companies had their own pathologists who carried out the tests as part of internal quality assurance programmes. Phytosanitary certificates were based on tests made by public laboratories such as the California Department of Agriculture. Field crop seed companies had very little concern about transmission of pathogens by seeds for domestic production, except for a few special cases such as loose smut of cereals and barley stripe mosaic virus. Seed health testing for these pathogens was handled by state

 Seed Health Testing (eds J.D. Hutchins and J.C. Reeves)

laboratories or universities. For the majority of field crop seeds, phytosanitary certificates were issued based on field inspection.

The seed health situation for field crops changed dramatically in 1988 with the expansion of markets for soybean and corn seeds in the European Community (EC). Laboratory tests were required for some of the most important pathogens. In one year, the seed health testing load at Iowa State University went from an annual average of 300 tests per year to 3000. Another significant change took place in 1994, with the implementation of a wide range of new phytosanitary regulations imposed by Mexico, the USA's biggest market for seeds. This has greatly diversified the demand for national seed health tests, to include many new crops and diseases. The Iowa State University Seed Science Centre now routinely runs almost 50 seed health tests compared with five in 1988.

Ideally the world phytosanitary system should protect against the spread of economically important pathogens without causing unnecessary barriers to the international movement of seeds. The world phytosanitary system has two main problems, however: the widespread use of unjustifiable phytosanitary regulations, and a lack of standardization of tests.

PHYTOSANITARY REGULATIONS

Unjustifiable regulations have had a serious economic impact on US seed exporters in recent years. All soybean seed lots exported from the USA to the EC have to be tested for soybean bacterial blight. This disease is widespread in the USA and is of no economic importance. It also has been known in Europe for many years. Since 1988, US seed companies have spent over $1,500,000 on testing for this pathogen on soybean seeds.

Two main reasons exist for these unjustifiable regulations: lack of access by regulators to scientific information on the pathogen in question, and trade barriers by which countries use phytosanitary regulations as a replacement for tariffs. Trade barriers are becoming more frequent as free trade zones, such as NAFTA, EC and the GATT agreement, develop. In recent years, the American Seed Trade Association (ASTA) has taken a proactive role in helping to contain the flood of phytosanitary regulations by utilizing the resources of the ASTA seedborne disease and phytosanitary committees to prepare documentation to challenge new regulations. Teams of experts also travel to the major trading partners to pursue these challenges. This approach has had significant impact on reducing some of the proposed regulations. The problem is becoming so overwhelming, however, that this approach is no longer adequate. Currently ASTA relies on pathologists within the industry to volunteer their time to seek out and organize the information to challenge regulations. These individuals, however, also have regular responsibilities in their employment, which limits the time they can give.

Challenges to phytosanitary regulations are severely handicapped without facts to justify a particular case. We all know how important intelligence is to a military operation or why a discovery phase is necessary for successful litigation. At present, our intelligence system is inadequate for dealing with phytosanitary issues. A new Seedborne Disease Database that is under construction can provide that system by allowing rapid and easy access to scientific information on seedborne disease. This should greatly improve the capacity of countries to challenge unreasonable regulations, and also improve the world phytosanitary system by enabling phytosanitary authorities to make more justifiable and rational decisions on regulations. This database is a joint effort of CABI (Centre for Agriculture and Biosciences International), the Danish Institute for Seed Pathology, and the Seed Science Centre, Iowa State University. The scope of the project will comprise economically important seedborne pathogens of all crops. Documented information will be provided for each pathogen on its economic importance, geographic distribution, significance of seedborne infection with respect to seed quality and transmissibility of the pathogen, seed treatment, and seed health assays. The first module of the database, designed primarily for south-east Asia, is scheduled to be produced as a CD-ROM disk in early 1997. A total of 150 seedborne disease data sheets will be included. The second, global phase will be available in 1998 that will comprise at least another 150 seedborne disease data sheets.

STANDARDIZED SEED HEALTH TESTS

The second major problem in the international phytosanitary system is the lack of standardized seed health tests. This has created serious problems for the seed industry in that the importing and exporting countries use different protocols to conduct a test and get different results. Seed companies then have to destroy or redeploy the seed lots at considerable expense. Lack of standardization is also a confounding factor when litigation ensues following transmission of seedborne diseases.

The ASTA has recently taken the lead in addressing the problem by the formation of a Seed Health Initiative Committee. This committee has developed a national plan to accredit both public and private laboratories nationwide to carry out seed health tests using standardized protocols. USDA-APHIS is now finalizing the plan and the rule will shortly be published in the Federal Register. The plan will be administered by an accreditation body appointed by APHIS. This body will ensure that applicant laboratories meet predetermined standards of facilities and equipment and expertise of personnel. To accomplish this, assessors would visit the facilities, staff would be sent to training sessions on standardized protocols, and annual proficiency tests would be arranged to ensure that the laboratories met quality standards. The project is requesting no funds from the industry or federal government, but will be supported by accreditation fees.

Concurrent with the US initiative, an international working group was formed to standardize seed health methods for vegetable crops. This initiative is known as the International Seed Health Initiative (ISHI) (see Chapter 12). This group has met in The Netherlands, France, USA, and UK since 1994. This project also is industry-initiated in response to serious economic problems imposed by seed transmitted pathogens. The primary mission of this effort is to establish protocols that will be recognized as standard tests for vegetable seed throughout the world. It is planned to bring field crop seeds into this group very shortly. This mission is carried out by cooperative efforts of pathologists in seed companies and in public state-supported agencies.

The final group involved in seed health testing is the International Seed Testing Association Plant Disease Committee (ISTA-PDC). For 30 years, the ISTA-PDC has produced Working Sheets for seedborne diseases that are a good model for standardized tests. At this time, efforts are being made to coordinate this activity with other national and international efforts to further standardize seed health tests.

If these initiatives continue to be pursued, an international phytosanitary system will be in place that is rational, effective and less expensive to both the seed industry and regulatory agencies worldwide.

Development of Seed Health Testing Policy in South America

16

J.C. Machado[1], M.G.G.C. Vieira[1], C.A. Coser[2] and L.A. Mendonça[2]

[1]*UFLA, C. Postal 37, Lavras, MG, CEP 37200-000;*
[2]*MARA/SDA/DDIV/CLAV, Brasilia, DF, Brazil*

INTRODUCTION

The history of seed pathology in South America could be described briefly, considering the few developments in this field compared with agricultural developments in developed countries. Although agricultural activity is of fundamental importance in South American countries for both food supply and economic development, the level of technology employed by farmers can be considered low and production of many crops has been gradually declining. Poor seed quality could be argued to be one of the most important contributory factors limiting the improvement of crop yields.

The incorporation of seed health testing into seed certification programmes has been a preoccupation in several South American countries in the past few years. Seed health testing has been introduced as a means to ensure seed quality as well as an effective measure to prevent introduction and dissemination of dangerous plant pathogens into previously uninfected areas. Implementation of such testing has, however, been slow and faced difficulties in terms of infrastructure and the low level of demand, by farmers, for high quality seed.

An estimate of the significance of agriculture in the South American continent can be gained from Table 16.1, which summarizes the geographical and climate characteristics of each country. The high potential of agricultural activity in this part of the world can be seen. The wide variation in climate and soil fertility gives the majority of South American countries the possibility of successfully cultivating a wide range of plant species propagated by seeds.

(eds J.D. Hutchins and J.C. Reeves)

Table 16.1. Geographical area, arable land and population of the South American countries.

Countries	Area (km^2)	Arable land (1000 ha)	Population (millions)
Argentina	2,766,889	25,000	34.6
Bolivia	1,098,581	2120	7.4
Brazil	8,547,403	49,500	155.8
Chile	756,629	3972	14.3
Colombia	1,141,748	3920	35.1
Ecuador	272,045	1633	11.5
French Guiana	91,000	10	1.1
Guyana	214,970	480	0.7
Paraguay	4 06,752	2190	5.0
Peru	1,285,261	3400	23.8
Surinam	163,265	57	0.4
Uruguay	176,215	1260	3.2
Venezuela	912,050	3215	21.8

Source: FAO/PYB (1993).

Crops such as soybean, cotton, maize, rice, sorghum, wheat, barley and many vegetables have been traditionally cultivated. For countries like Brazil and Argentina, agriculture has formed a vital and substantial component of their economy, unlike most other countries in the continent (FAO/PYB, 1993; Reis, 1996). There is a common belief that approximately 25% of the whole area of South America is considered suitable for cultivation of arable crops. Around 40% of the total area is occupied by natural forests (Amazon) and artificial forests. Fifty per cent of the total land area is composed of savana, a natural vegetation also named *Cerrado*, characterized by acid soils with low levels of plant nutrients (Lopes and Guilherme, 1991).

It has been estimated that, in South America, close to 50 million hectares are cultivated annually with arable crops. This equates to the production of approximately 130 million tons of grain and an annual demand of 5 million tons of seed. Because of the level of cultivation technology used by farmers in this continent, it can be assumed that crop productivity remains low compared with the crops' genetic potential, as demonstrated through experimental plots in different areas of arable land in the continent (Lopes and Guilherme, 1991).

According to Reis (1996), the South America Common Market (Mercosul) is generally considered to be one of the largest producers of agricultural commodities in the world (Table 16.2). The level of agricultural productivity in South America has generally not achieved its considerable potential. The major factor limiting this development is the low investment in fundamental and applied research into the agricultural sciences in addition to the non-existence of a sound and permanent policy for this kind of research.

Table 16.2. Grain production of the major seed propagated crops in the Mercosul countries in 1994–95.

	Production (tons × 1000)				
Crop	Argentina	Brazil	Paraguay	Uruguay	Total
Soybean	11,933	25,934	1680	18	39,565
Corn	10,439	34,208	500	–	45,147
Rice	545	11,243	–	680	12,468
Wheat	9223	1600	320	300	11,443
Cotton	450	565	130	–	1145

Source: CONAB/FARSUL (1995).

SEED TECHNOLOGY AND CERTIFICATION SCHEMES IN SOUTH AMERICA

The lack of reliable records, and the poor integration and cooperation between institutions, scientists and seed producers in South America, ensures that information regarding seed certification is hard to collate. The facts on which this paper was based are the result of replies from only a limited number of colleagues. Some of the examples presented are from the views and experience of the authors.

In general, it can be assumed that the existing seed certification schemes in South American countries follow a similar model of other country members of the International Seed Testing Association (ISTA). However, modifications to some of the rules, the certification seed classes and to the certification programme legislation depend upon the technological level of each country and the growing season. Essentially, the certification scheme controls the genetic purity and quality of all aspects of seed production in the field, during harvest, processing, storage and finally inspection in the market. The certification programme starts with the multiplication of the foundation/basic seed, which comes from the plant breeders, and ends with the production of certified seed, which can be further classified into two or three categories (Fig. 16.1). Seeds are distributed to farmers under the guarantee of quality, in terms of genetic and physical purity and germination capacity, as indicated in the certification bulletin issued by accredited laboratories. Normally the system is controlled or coordinated by central government with the participation of local agricultural authorities. In some countries, specific and official organizations like the National Institute for Seed Analysis (INASE) in Argentina and the Central Laboratory of Plant Analysis (CLAV) in Brazil exist to take care of the seed programmes in general. Multiplication of seed is carried out by qualified farmers who may work either alone or in cooperatives. In Brazil, for example,

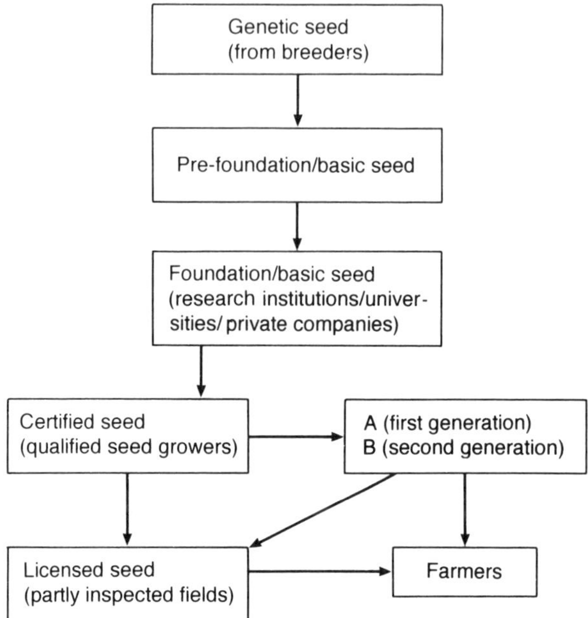

Fig. 16.1. Schematic representation of the seed certification programmes followed in South America.

multiplication of foundation seed is the task of the National Enterprise of Agricultural Research (EMBRAPA), which in some cases involves the direct participation of official and private institutions and selected farmers.

Field inspection is an important phase of the certification scheme and is conducted by official inspectors; it consists of the evaluation of the genetic characteristics of the cultivars or breeding lines, occurrence of diseases, weeds and other aspects regarding the management of the crop. For each crop, standards (Table 16.3) and numbers of inspections are established. Standards may vary between states or provinces, but they cannot exceed the national settings.

Certification of seed quality is based on rules described in handbooks, which follow the International Rules published by ISTA as closely as possible. Regardless of the existence of certification programmes for most species cultivated in the South American continent, a considerable number of farmers still use home-saved seed instead of certified seed. In Mercosul countries (Brazil, Argentina, Uruguay, Paraguay and Chile) the need to increase and improve yields for export or domestic demand has led to farmers demanding higher quality seeds.

Table 16.3. Examples of selected field standards established in certification programmes for plant diseases in some South America countries.

Pathogen/host	Seed class	Tolerance level (%)	Reference
Cotton			
Fusarium wilt	B, C1, C2	0	Lobato *et al.* (1985)
	B	0.2[b]	Mehta (1986)
	C	0.5[b]	Mehta (1986)
Phaseolus			
Common bacterial blight	B, C1	0.5[a]	Lobato *et al.* (1985)
	B, C	0[b]	Yorinori (1988)
	C2	1.5[a]	Lobato *et al.* (1985)
	C	0[c]	C.A. Silva, personal communication
Anthracnose	B, C1	0.5	Lobato *et al.* (1985)
	B, C	0	Yorinori (1988)
	C2	4.0 (on pods)	Lobato *et al.* (1985)
	C	0[c]	C.A. Silva, personal communication
Wheat[d]			
Smut (*Ustilago* spp.)	B	500 plants ha^{-1}	COSAVE (1995)
	C	1000 plants ha^{-1}	COSAVE (1995)
Bunt (*Tilletia* spp.)	B, C	50 plants ha^{-1}	COSAVE (1995)

Abbreviations: B, basic seeds; C, certified seeds (C1, first generation; C2, second generation).
[a]Standards for the State of Minas Gerais, Brazil.
[b]Standards for the State of Paraná, Brazil.
[c]Standards in Colombia.
[d]Standards in Argentina.

THE DEVELOPMENT OF SEED HEALTH TESTING IN SOUTH AMERICA

Recognition of the importance of seed health in South America began in the mid-1970s (Yorinori *et al.*, 1979; Wetzel *et al.*, 1981). Before then, a considerable number of reports can be found in the literature mostly dealing with seed treatment and records of microorganisms in seed lots (Wetzel *et al.*, 1981). For certification purposes, seedborne diseases had been only considered at the field stage with little attention to the importance of seed infection.

Implementation of routine seed health testing, in some countries of the continent, has been an aim for 15 years. The main objective has been to incorporate standard seed health testing methods and tolerance levels for important seedborne pathogens into the certification programmes. This was considered to be of prime importance to improve the seed health quality of the most important agricultural crops. An example of this can be provided by a strategy adopted in Brazil, which resulted from three meetings. From the first meeting, a Latin America Workshop held in Londrina in 1975, a National Program of Seed Pathology was launched. From the two other meetings – the First Brazilian Workshop on Seed Quality Control held in Lavras in 1991 and the National Congress of the Brazilian Seed Technology Association (ABRATES), held in Campo Grande in the same year – issues were formulated proposing standards for several seed-transmitted pathogens, selected examples of which are highlighted in Table 16.4. Implementation of this programme should be initiated by submitting all foundation seed lots for seed health testing, during an experimental period of 2–3 years. Seed treatment is recommended to control all pathogens occuring at any level up to the standard indicated in the programme.

Table 16.4. Proposed seed health standards (%) for selected pathogens of some cultivated crops to be adopted during a 2–3 year experiment in Brazil.

	Seed certification class	
Crop/pathogen	Basic	Certified
Cotton		
Colletotrichum gossypii	20.0	30.0
Xanthomonas campestris pv. *malvacearum*	0	1.0
Common bean (Phaseolus)		
Colletotrichum lindemuthianum	0	1.0
Phaeoisariopsis griseola	2.0	3.0
Sclerotinia sclerotiorum	0	0
Xanthomonas campestris pv. *phaseoli*	0	0
Maize		
Cephalosporium acremonium	30.0	40.0
Colletotrichum graminicola	2.0	5.0
Drechslera maydis	5.0	7.0
Fusarium moniliforme	50.0	60.0
Peronosclerospora sorghi	0	0
Rice		
Drechslera oryzae	15.0	20.0
Pyricularia oryzae	15.0	20.0
Soybean		
Cercospora kikuchii	25.0	30.0
Cercospora sojina	5.0	10.0
Phomopsis spp.	35.0	40.0
Sclerotinia sclerotiorum	0	0
Wheat		
Bipolaris sorokiniana	30.0	35.0
Xanthomonas campestris pv. *undulosa*	0	0

From Machado (1994) with modifications proposed by the Committee of Seed Pathology/ABRATES in 1995 and MARA in 1996.

Another important action taken in Brazil and Argentina over the past few years has been the accreditation of seed pathology laboratories for performing seed health tests. At present, 21 laboratories in Brazil are qualified and accredited to perform seed health tests for the detection of pathogenic fungi of the common crops, following methods described in the Brazilian rules for seed analysis (Anon., 1992). Twenty seed analysis laboratories in Argentina have also been reported to be qualified to perform seed health tests following methods in the ISTA Handbook (INASE, personal communication; FAO, 1993).

Although several methods are described in the Brazilian rules for seed analysis (Anon., 1992), the blotter test and rolled paper towel test are the most commonly used methods in routine analysis for detection of seedborne fungi. A more recent programme organized by the Brazilian Ministry of Agriculture had the objective of surveying the health status of certified seed produced in different regions in the country during a 3 year period (1995–1997). Following this survey, seed health testing is expected to become compulsory for some crop–pathogen combinations.

To describe the current stage of development of seed health policy in South America, it should be emphasized that there are ongoing discussions between member countries of Mercosul, aiming to set up common seed health legislation for some crop species. As a result of several meetings over the last 5 years, seed health standards have been proposed for pathogens of two species: wheat and maize. Establishment of standards for pathogens in onion, garlic and tomato seeds has been under discussion and is shortly expected to be defined. The definition of methods, standards and common procedures for other crops are expected to be discussed in a workshop which has been organized for 1997.

CONCLUSION

The integration of South American countries as a result of world trade globalization can be considered as one of the main reasons for an expected rapid development of seed health policy in this continent. The need to produce high quality seeds is now making governments and private companies look at this aspect more closely. In that respect development of modern methodologies for more precise and accurate detection of seedborne pathogens and seed treatment to control the diseases caused by these pathogens, have been topics under investigation in several countries. In addition to improving conventional methods for detecting certain pathogens in routine analysis, more specific and precise methods employing molecular techniques are under investigation for the detection of some of the most economically important seedborne pathogens. Emphasis has been given to fungi, such as *Colletotrichum gossypii* var. *cephalosporioides*, *Sclerotinia sclerotiorum* in several species, *Diaporthe* sp. in soybean (Jaccoud and Reeves, 1993; Nasser *et al.*, 1996; Peres *et al.*, 1996; Vieira *et al.*, 1996).

The important role played by the Seed Pathology Committee of the ABRATES in implementing a seed health testing programme in Brazil over the past 12 years should also be recognized. During this period four national symposia, two Latin America workshops and several short training courses, including one for Latin America countries with the participation of ISTA–PDC instructors, have been held. As the result of this programme two seed pathology textbooks have been published in Portuguese (Soave and Wetzel, 1987; Machado, 1988). Finally it should be stated that although it has been recognized as an important tool in the process to increase food production, implementation of seed pathology in the seed certification and plant protection programmes is a great challenge, requiring determination by governments in each country, and will take time to become consolidated.

REFERENCES

Anonymous (1992) Ministério da Agricultura *Regras para Análise de Sementes.* Brasília SNAD, Departamento Nacional de Defesa Vegetal, CLAV, 365 pp.

CONAB/FARSUL (1995). In: *Relatório da Gazeta Mercantil*, p. 36.

COSAVE (Comitê de Sanidade Vegetal do Cone Sul) (1995) Standard sobre certificação sanitária de trigo. In: *Anais do IV Reunião do Grupo de Trabalho Permanente em Sanidade de Materiais de Propagação Vegetal.* Porto Alegre, Brasil, Anexo III, pp. 6.

FAO (1993) *Integracion y Normalization de los Metodos de Analisis de Semillas.* Programa de Cooperacion Tecnica, Document 1. FAO, Rome, 75 pp.

FAO/PYB (1993) *Year Book.* Food and Agriculture Organization of the United Nations v.47.

Jaccoud, D.S. and Reeves, J.C. (1993) Detection and identification of *Phomopsis* species in soyabean seeds using PCR. In: Sheppard, J.W. (ed.), *Proceedings of the First ISTA–Plant Disease Committee Symposium*, Ottawa, Canada, pp. 34–43.

Lobato, L.C., Selma, M., de Carvalho, J.R.M. and Carvalho, V.M. (1985) *Normas, padrões e procedimentos para a produção de sementes básicas, certificadas e fiscalizadas.* Belo Horizonte, Brasil, pp. 5–110.

Lopes, A.S. and Guilherme, L.R.G. (1991) *Preservação ambiental e produção de alimentos.* Associação Nacional para Difusão de Adubos e Corretivos Agrícolas. São Paulo, 16 pp.

Machado, J.C. (1988) *Patologia de Sementes: Fundamentos e Aplicações.* Ministério da Educação, Brasília, ESAL/FAEPE, Lavras, 107 pp.

Machado, J.C. (1994) Padrões de tolerância de patógenos associados à sementes. In: W.C. Luz. (ed.), *Revisão Anual de Patologia de Plantas*, Vol. 2. Passo Fundo, pp. 229–263.

Mehta, Y.R. (1986) Estabelecimento de padrões de tolerância para sanidade no campo e na semente. In: *Simpósio Brasileiro de Patologia de Sementes*, Vol. 2. Campinas, Anais; Fundação Cargill, Campinas, pp. 41–56.

Nasser, L.C.B., Boland, G.J., Sutton, J.C., Ferraz, L.C.L., Perez, A.P., Machado, J.C. and Care Filho, A.C. (1996) A new approach to detect *Sclerotinia sclerotiorum* on seeds. In: *Proceedings of Second ISTA–Plant Disease Committee Symposium*. National Institute of Agricultural Botany, Cambridge.

Peres, A.P., Silva, R., Vieira, M.G.G.C., Machado, J.C. and Santos, J.B. (1996) Analysis of genetic diversity of some isolates of *Sclerotinia sclerotiorum* using RAPD markers. In: *XV Seminário Panamericano de Semillas III Workshop sobre Marketing em Sementes e Mudas*, Gramado, Brazil, pp. 156.

Reis, R.P. (1996) *Integração Econômica e o Mercosul*. Lavras, Brazil, pp. 1–22.

Soave, J. and Wetzel, M.V.S. (1987) *Patologia de Sementes*. Fundação Cargill, Campinas, Brazil, 480 pp.

Vieira, M.G.G.C., Machado, J.C. and Santos, J.B. (1996) Differentiation between *Colletotrichum gossypii* south and *Colletotrichum gossypii* var. *cephalosporioides* by using RAPD markers. In: *Proceedings of Second ISTA–Plant Disease Committee Symposium*. National Institute of Agricultural Botany, Cambridge.

Wetzel, M.M.V.S., Bettiol, E.M. and Faiad, M.G.R. (1981) *Bibliografia Brasileira de Patologia de Sementes*. EMBRAPA/CENARGEN, Brasília, 256 pp.

Yorinori, J.T. (1979) Data showing results of comparative seed health tests of soybean. Discussion on results of comparative seed health tests of soybean. In: Yorinori, J.T., Sinclair, J.B., Mehta, Y.R. and Mohan, J.K. (eds), *Seed Pathology Problems and Progress*. IAPAR, Paraná, Brazil, pp. 42–45.

Yorinori, J.T. (1988) Importância do aspecto sanitário em programas de produção de sementes. In: Machado, J.C., Vieira, M.G.G.C. and Silveira, J.F. (eds), Anais do 3 *Simpósio Brasileiro de Patologia de Sementes*, Vol 3. Lavras, 1988. Fundação Cargill, Campinas, Brazil, pp. 29–47.

Developments in Rice Seed Health Testing Policy

17

T.W. Mew

International Rice Research Institute, PO Box 933, Manila 1099, The Philippines

INTRODUCTION

Seed health concerns the quality status of the seed which may be used either as farmers' planting material or as germplasm for crop improvement or research. Seed health is also targeted for crop improvement and production. For crop improvement and research, seeds are often moved across international boundaries. The health status is significant in the sense that organisms carried by seeds may be moved to an area or country where the organisms have not previously been introduced. Plant quarantine regulations have been developed to guide the movement or introduction of such germplasm.

Seed can also be a commodity of trade. Seed health testing requirements for seed as a commercial product or for scientific purposes, even of the same crop, are treated differently. In general, a country may prohibit the importation of foreign seed because of the potential threat of introducing pests or pathogens. The prohibition applies only to seed imported as a commercial product. Seed imported for scientific purposes may be exempt from the restriction. Access to such seed is restricted and assessment of the hazards and risks is easily performed. Persons who import these seeds have to follow general plant quarantine guidelines. Justification for this double standard is the favourable risk/benefit ratio for scientific material and the much less favourable ratio for commercial material – when agricultural production of the country as a whole is considered (Kahn, 1986). This discussion of rice seed health testing is limited to the use of seed for scientific purposes only. Although the objectives are different, the testing procedures are applicable to seed as commercial product of international trade.

THE CONCERNS OF SEED HEALTH TESTING

The detrimental effects of plant pests (pathogens, insects and weeds) on seed quality and seed health are generally well understood but seldom considered in actual crop production, especially at the farmer level in developing countries. Seed health is very important in sustainable crop production. It is the starting point of obtaining a high yield. Correct management of the crop should then secure the genetic potential of high yielding varieties. Information based on seed health testing can be applied to farmers' seed management practices for crop production. Seed health is a highly significant aspect of agricultural production and seed health testing provides a means to realize such potential. Seed health may be considered from two viewpoints: qualitatively (whether seed is infected with a plant pathogen) and quantitatively (the level of such infection or contamination). There is a need to establish the quantitative aspects in setting the tolerance level for the presence of certain lower-risk pests and pathogens.

SEED HEALTH FOR SUSTAINABLE CROP PRODUCTION

Experience in south-east Asia has shown that farmers often change their rice varieties in their crop production after several seasons of use. Lower productivity may be due to the poor seed health status after several seasons since farmers usually save some of their seed for subsequent plantings. In general, governments in the region can only supply a proportion of certified seed required for planting. In Malaysia this equates to approximately 30% of the seed required (Hamid, 1996). Farmers' seed management practices are often inadequate to ensure their home-saved seed is of a high quality.

Use of clean healthy seed is an important prerequisite for good crop establishment and production. Certified seed ensures genetic purity and freedom from grassy weed infestation and common seedborne pathogens. In Malaysia farmers are advised to use their own seeds for planting in order to complement the shortage of certified seed (Hamid, 1996). In Thailand, most farmers change their rice varieties every three to five crops or in 3 year intervals due to 'red rice' seed contamination (Somkid *et al.*, 1996). In Vietnam more than 60% of the farmers change their rice variety every season each year. It has been reported that only 30% of the farmers use the same varieties for more than 2 years (Khoa *et al.*, 1996). Between 20 and 70% of farmers choose to use their home-saved seed. The remainder obtain seed either through exchange among themselves or buy seeds from other farmers or relatives. Few farmers buy seed from seed growers or seed agents. The reasons why farmers change their varieties include the wish to use the latest varieties, and concerns that seed quality has deteriorated and that continuous planting of the same varieties for more than one season could lead to yield decline or induced pest susceptibility. Preliminary results from a survey indicated that by improving the health status of their home-saved seed, farmers could easily obtain a 10–20% increase in yield (Merca and Mew, unpublished data). This improvement was achieved by winnowing, flotation and physical sorting to remove discoloured seed, mixtures and weed seeds.

IRRI'S ACTIVITY WITH RICE SEED HEALTH TESTING

It has long been recognized that pests, weeds and pathogens are carried with rice seeds as rice is transported between countries and continents. To maintain germplasm in a seed bank the seed must be in good health when it is stored. Seed health testing is carried out to ensure this. The International Rice Research Institute (IRRI) recognizes two important aspects of seed health testing: safe germplasm exchange and pest management. Quality seed is not only a plant quarantine requirement for germplasm movement, but it is also an essential step in crop protection to produce a good crop and to realize the genetic potential of improved crop cultivars.

The approach for rice seed health testing taken at IRRI involves tests on both incoming and outgoing seed lots. The procedures include crop health inspection, dry seed inspection, routine seed health testing and seed treatment (Mew and Merca, 1992). The principle is to ensure that only seeds of the highest quality are sent to consignees. Only pure seed, dried to about 12% moisture content and high in germination potential, is certified for shipment.

At IRRI, to integrate seed health testing into producing good quality seed for crop production, seed is classified into four categories based on the physical conditions of the seed. The best (Category 1) seeds are clean, unblemished, healthy-looking, well developed, without deformity, having no sign of discoloration and are free from fungal infection. Seed in Categories 2–4 needs different levels of quality improvement by sorting and cleaning. Macroscopic examination is not conducted for seeds in these latter categories until the contaminants have been removed. Seed lots containing soil particles, seed mixtures, insects and sign of insect infestation, weed seeds, smutted seed, sclerotia and other plant parts require extensive cleaning before further testing. For incoming seed, following the Association of South-East Asian Nations (ASEAN) Plant Quarantine Regulations, a seed lot may be rejected if it has more than 20% contamination or infestation, or fungal infection such as caused by *Tilletia barclayana* (see Chapter 20, for method of detection).

The seeds released by IRRI's Seed Health Unit to users outside The Philippines are always accompanied by a phytosanitary certificate issued by an

Table 17.1. Frequency of detection of selected fungal pathogens associated from untreated incoming rice seed lots (1991–95).

	Detection frequency (%)[a]					
Pathogens	1991	1992	1993	1994	1995	Mean
Curvularia spp.	91.7	89.9	97.5	90.0	92.5	92.3
Alternaria padwickii	93.9	93.1	96.1	85.9	89.6	91.7
Sarocladium oryzae	27.8	50.9	57.4	15.2	28.9	36.0
Fusarium moniliforme	39.6	38.4	14.0	28.8	32.8	30.7
Bipolaris oryzae	77.4	55.4	75.4	74.2	49.3	66.3
Tilletia barclayana	14.4	17.0	12.0	4.6	11.0	11.8
Pyricularia oryzae	3.5	10.7	2.0	1.6	11.9	5.9

[a]Based on a total of 26,755 seed lots.

authorized Plant Quarantine Officer based on the seed health testing results conducted at the IRRI. Recently, shipments have also been accompanied by a health status declaration authorized by the Head of the Seed Health Unit on behalf of IRRI, to indicate that the test has been conducted at the IRRI and that we are willing to provide the test record if it is needed.

Routine seed health testing is carried out for a range of plant pathogens (Tables 17.1 and 17.2). Samples are drawn from each seed lot following the sampling method prescribed by the ISTA Rules (ISTA, 1985). The blotter and growing-on test methods are used in the laboratory (to detect fungi and bacteria respectively). As no seedborne viruses of rice are known, virus tests are not conducted.

CRITERIA DETERMINING NEED FOR QUARANTINE

Besides plant pathogens rice seed is often contaminated with other objects including weed seeds, insect infestation (mostly storage insects), soil particles, seed with soil particles, smut or smutted seeds, any of which would lead to quarantine of the seed. For international seed exchange or movement, seed contamination is only part of the basis of prohibition or rejection of importation. There are other issues often neglected by import request (Kahn, 1982): (i) contamination, infection or infestion with pests or pathogens of quarantine significance against which there is no known practical and effective eradicant treatment that can be administered safely at the port of entry; (ii) mixture with or contamination by prohibited species; (iii) contamination with prohibited articles such as straw, soil or bark; (iv) lack of a permit issued by the importing country or a phytosanitary certificate issued by the exporting country or both; (v) lack of the required added declarations on the phytosanitary certificate in accordance with the conditions stated on the permit; and (vi) having a phytosanitary certificate that has been erased or altered.

Table 17.2. Levels of infection caused by selected fungal pathogens detected on rice seed of untreated income seed lots (1991–95).

	Level of infection (%)[a]					
Pathogens	1991	1992	1993	1994	1995	Mean
Curvularia spp.	8.3	17.5	15.0	7.5	10.8	11.8
Alternaria padwickii	22.9	10.1	14.8	9.7	19.3	15.4
Sarocladium oryzae	2.0	2.7	4.9	2.0	4.0	3.1
Fusarium moniliforme	1.2	3.5	2.0	1.4	1.7	2.0
Bipolaris oryzae	5.8	7.2	3.5	13.9	6.4	7.4
Tilletia barclayana	42.0	14.1	10.4	17.8	13.0	19.5
Pyricularia oryzae	2.8	2.1	1.0	6.2	3.8	3.2

[a]Based on a total of 26,755 seed lots.

In rice the first three of these items are directly related to seed health testing and are under our control. According to Kahn (1982), prior to the 1980s, rice was listed as a crop frequently prohibited by countries in relation to its health status and 79% of these cases involved seed imports. In the past decade, there have been very few incidences where rice is still a prohibited crop. We believe this is partly attributed to the general awareness of the importance of rice seed health testing. Experience at the IRRI suggests that in the few cases where rice seed was prohibited by The Philippines, they were all related to hybrid rice seeds with high level of infection by *T. barclayana*.

Prohibition is the most drastic action taken by plant quarantine services. There should be sound biological justification for exercising this regulatory action in view of the potential economic, political and biological impact of phytosanitary exclusion. Prohibition should be exercised: (i) when the pest or pathogen does not occur in the importing country or, if it does occur, it is not widely distributed; and (ii) when the pest or pathogen causes economic damage, or has the potential to cause such damage, the plant or seed cannot be inspected for an unidentified pest or pathogen at port of entry by a plant quarantine officer using equipment and procedures found routinely in a plant inspection station, or it can be inspected but there is no practical, safe and effective eradicant treatment that can be administered should a pest or pathogen meeting those criteria be encountered (Kahn, 1988).

DETECTION FREQUENCY OF SEEDBORNE PATHOGENS AND SEED INFECTION LEVELS

The purpose of seed health testing is to detect seedborne pathogens; however, it is not always possible to eliminate all possible seedborne pathogens. Rice is host to five major groups of plant pathogens, many of which are seedborne. These

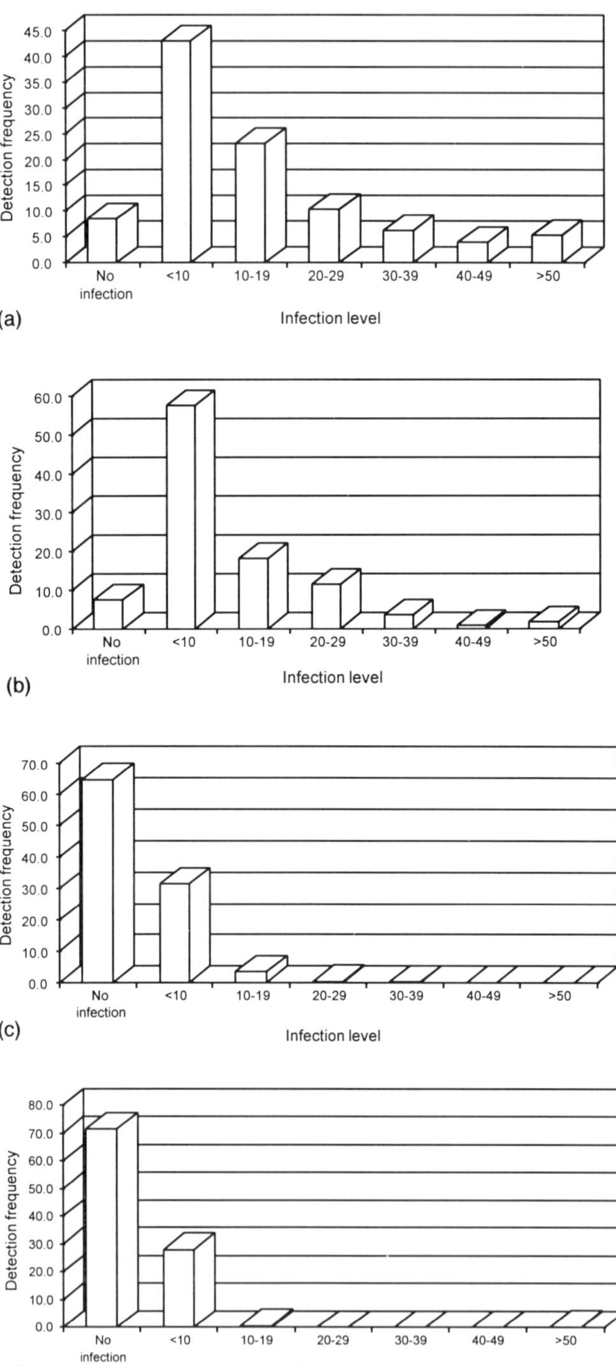
45.0
40.0
35.0
30.0
25.0
20.0
15.0
10.0
5.0
0.0
Detection frequency
No infection
<10
10-19
20-29
30-39
40-49
>50
(a)
Infection level
60.0
50.0
40.0
30.0
20.0
10.0
0.0
Detection frequency
No infection
<10
10-19
20-29
30-39
40-49
>50
(b)
Infection level
70.0
60.0
50.0
40.0
30.0
20.0
10.0
0.0
Detection frequency
No infection
<10
10-19
20-29
30-39
40-49
>50
(c)
Infection level
80.0
70.0
60.0
50.0
40.0
30.0
20.0
10.0
0.0
Detection frequency
No infection
<10
10-19
20-29
30-39
40-49
>50
(d)
Infection level

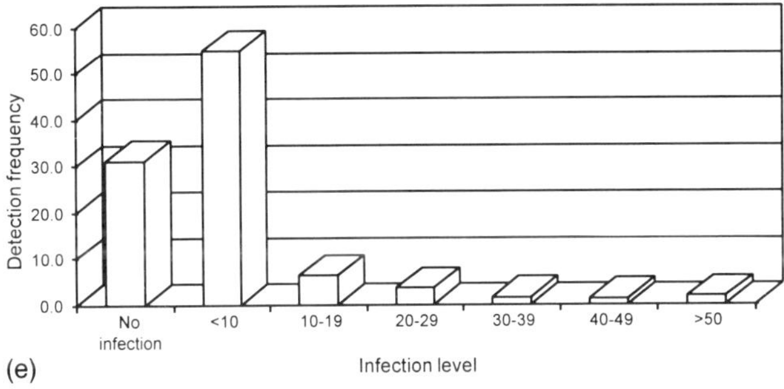

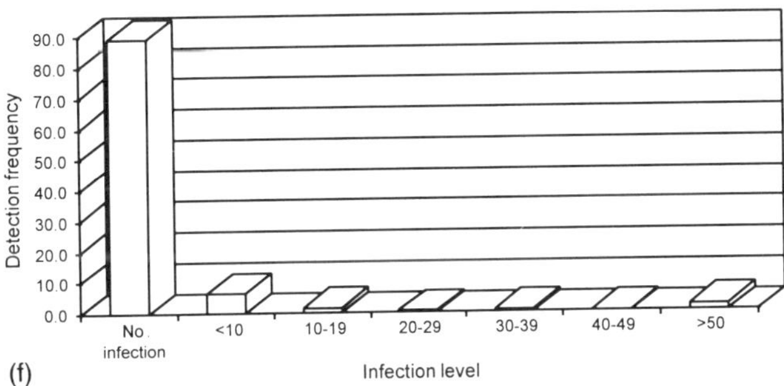

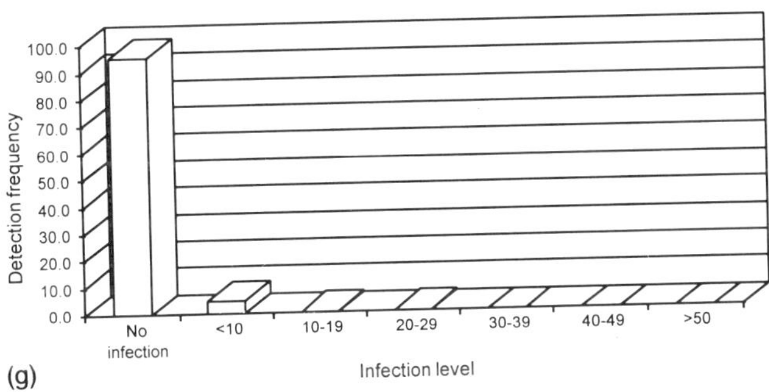

Fig. 17.1. Frequency of detection of (a) *Alternaria padwickii*, (b) *Curvularia* spp., (c) *Sarocladium oryzae*, (d) *Fusarium moniliforme*, (e) *Bipolaris oryzae*, (f) *Tilletia barclayana* and (g) *Pyricularia grisea* in 26,755 incoming seedlots in 1991–95.

pathogens are also worldwide in distribution (Ou, 1985; Mew *et al.*, 1988). Whether or not the seedborne inoculum plays an important role in disease epidemics depends on the type of disease, the rice production environment and farmers' cultural practices.

At the IRRI, we have conducted seed health tests, for phytosanitary certification purposes, from all rice growing countries over the past 20 years. Each year, we have tested between a few thousand and 50,000 seed lots. Using the data from 1991–95 of incoming seed lots, we established the detection frequency and infection levels of some of the known seedborne fungal pathogens. A few selected rice fungal pathogens are included in the present discussion. The operational definition of detection frequency is based on the detection of a pathogen of a seed lot, while infection level defines the percentage of the seeds infected within a seed lot. From the data, it is clear that both the detection frequency and infection level are very high for *Alternaria padwickii* (Fig. 17.1a), followed by *Curvularia* spp. (Fig. 17.1b). In rice production in tropical Asia, none of these organisms cause any noticeable damage to the crop in the field. The detection frequency of *Sarocladium oryzae* (Fig. 17.1c), *Fusarium moniliforme* (Fig. 17.1d), and *Bipolaris oryzae* (Fig. 17.1e) are moderate and *T. barclayana* (Fig. 17.1f) is low. Although *Pyricularia oryzae* (Fig. 17.1g), the rice blast pathogen, is considered to be a very important rice pathogen, the detection frequency as well as the infection level are the lowest among the pathogens so far detected. Except for *F. moniliforme*, seedborne inoculum of the other pathogens may not be the most important source of inoculum for these diseases (Mew *et al.*, 1988).

RISK ANALYSIS OF SEEDBORNE PATHOGENS

Since quarantine risks have not often been quantified, quarantine officers generally refer to the level of risks as low, medium or high. Evidently risk associated with rice seed in international exchange and movement is a perceived rather than an actual risk. As when often dealing with the unknown, it is normal to take a more conservative approach. We need to develop methods to analyse the actual risk of rice seedborne pathogens or inoculum for disease development. To date there is very little information in this aspect.

Recently Mew and Savary (1993) proposed methodology for analysing the risk associated with rice seedborne pathogens based on routine seed health testing data. Risk (R) is a function of risk probability and risk magnitude:

$$R = PM = ieM$$

where P is the risk probability; i is the introduction risk, i.e. the probability that a pathogen may enter an area (based on routine seed health testing data); e is the epidemic risk, i.e. the probability that this pathogen may establish infection from seedborne inoculum (obtained from experimental data of the pathogen under different rice production environments); and *M* is the risk magnitude, i.e. the potential consequence of an epidemic due to the introduced pathogen from seed (obtained from databases or modelling of crop loss). In rice, the crop loss database is available in a few selected environments.

Risk analysis should take into account the following: the type of pathogens, the role of seed in its life cycle, the disease or epidemic potential, the types of seedbed or nursery, or methods of planting, the rice production environment and the density and kinds of beneficial microorganisms, especially those demonstrated to have biological control ability. At present pest risk analysis for plant quarantine purpose is a process whereby the entry status of seed is determined based on perceived hazards. We do not have sufficient data (biological data and yield loss data) to provide plant quarantine with a definite conclusion based on seed health testing data.

CONCLUSION

At the IRRI, we are currently assessing the impact of rice seed health, the pathogens and non-pathogens associated with rice, in rice production. We believe such information is vital to establish the risk probability of the seed-borne pathogens through seed health testing, and to determine the impact of seed health on rice production. It may be possible that when the methodology is developed and crop loss databases are available then the usefulness of seed health testing may be broadened. For rice, we are in the developmental stage of all these.

REFERENCES

Hamid, A.A. (1996) Farmers' practices in rice seed production in Malaysian. In: *Proceedings of Seed Health for Disease Management Conference.* IRRI, Los Banos, Philippines, pp. 19–23.

ISTA (1985) International rules for seed testing. *Seed Science and Technology* 13, 299–355.

Kahn, R.P. (1982) Phytosanitary aspects of the international exchange of plant germplasm. In: *Proceedings of Plant Protection in the Tropics Conference.* MAPPS, Kuala Lumpur, pp. 101–116.

Kahn, R.P. (1986) *International Exchange of Plant Germplasm into and within the Caribbean Region.* Bulletin 3, North American Plant Protection Organization

Kahn, R.P. (1988) The importance of seed health in international seed exchange. In: *Proceedings of Rice Seed Health Conference.* IRRI, Loas Banos, Philippines, pp. 7–20.

Khoa, N.T., Lan, P.T.P., Dau, H.X., Dinh, H.D., Hach, C.A., Luat, V.N. and Paris, T.R. (1996) In: *Proceedings of Seed Health for Disease Management Conference.* IRRI, Los Banos, Philippines, pp. 26–29.

Mew, T.W. and Merca, S.D. (1992) Detection frequency of fungi and nematode pathogens from rice seed. In: Manalo, P.L., Chan, K.C. and Salayo, N.D. (eds), *Plant Quarantine in the '90s and Beyond*. ASEAN Planti, Kuala Lumpur, pp. 193–206.

Mew, T.W. and Savary, S. (1993) Risk analysis of rice seed-borne pathogens. In: *Sixth International Congress of Plant Pathology*, Montreal, Canada, Abstract S19.4, p. 21.

Mew, T.W., Bridge, J., Hibino, H., Bonman, J.M. and Merca, S.D. (1988) Rice pathogens of quarantine importance. In: *Rice Seed Health*. IRRI, Los Banos, Philippines, pp. 101–115.

Ou, S.H. (1985) *Rice Diseases*, 2nd edn. Commonwealth Mycological Institute, Kew, 380 pp.

Somkid D.S., Kamolsak K., Somsak T. and Kannika, P. (1996) Farmers' clean seed practices and seed health testing in Thailand. In: *Proceedings of Seed Health for Disease Management Conference*. IRRI, Los Banos, Philippines, pp. 23–26.

Development of a PCR-based Test to Detect and Identify *Pyrenophora* spp.

18

E.A. Stevens, E.J.A. Blakemore and J.C. Reeves

National Institute of Agricultural Botany, Huntingdon Road, Cambridge CB3 0LE, UK

INTRODUCTION

Pyrenophora graminea Ito & Kuribay (imperfect state *Drechslera graminea* (Rabenh.) Shoem.), *Pyrenophora teres* Drechsl. (imperfect state *Drechslera teres* (Sacc.) Shoem.) and *Pyrenophora teres* (Drechsl. f. sp. *maculata* Smedeg.) are the fungal agents of leaf stripe and net and spot blotch diseases on barley, respectively. These three pathogens are capable of causing serious crop losses in most countries where barley is grown. Glasshouse experiments have shown that the cumulative effect of successive inoculations of *P. teres* on winter barley resulted in a 58% reduction in yield (Jordan *et al.*, 1985).

Transmission of leaf stripe is primarily seedborne whereas net and spot blotch may also be dispersed via crop debris. It is important to be able to distinguish between these three pathogens for farmers to make appropriate management decisions for disease control. The most effective method of controlling seedborne diseases is through seed certification. Although *Pyrenophora* spp. are not directly covered by the seed certification procedures this disease complex is of considerable importance in the UK and voluntary standards for seedborne disease have been adopted (Cockerell *et al.*, 1995). Organomercury fungicides were used to control these diseases until their use was banned in 1992. With increasingly tight regulation of pesticide usage, it seems probable that in the long term the most effective means of seedborne disease control will be through seed health testing, seed treatment and the introduction of new resistant barley cultivars.

Traditional seed health tests involve growing the pathogen from seed plated out on agar or on moist filter paper (Rennie and Tomlin, 1984). These tests have several disadvantages: (i) a long incubation time is required for the pathogen to grow (7 days); (ii) the test is time consuming as identification of each of the pathogens must be carried out by microscopic analysis; (iii) tests may be insensitive as low levels of infection may not result in enough visible characteristic phenotypic features; and (iv) tests require highly skilled technicians who have had taxonomic training and are competent to identify each pathogen correctly.

The disadvantages of traditional seed health testing methods have led to much interest in the development of DNA techniques, which have the potential advantages of greater speed, sensitivity and specificity, as well as being more amenable to automation. Identification and detection methods employing the polymerase chain reaction (PCR) have been successfully used elsewhere to identify seedborne organisms (Blakemore and Reeves, 1993; Jaccoud and Reeves, 1993) as a means of improving seed health testing.

The objective of this research is to obtain specific primers for each of the three pathogens and subsequently to develop a multiplex PCR test (for example as given in Nicholson *et al.*, 1996). This would make it possible to have the normal advantages of specific PCR (i.e. sensitivity and specificity) with the additional capacity to detect more than one pathogen in a single test, thereby making the seed health test more cost-effective and efficient in its use of both resources and time. Two strategies were used to obtain specific primers for detecting the three *Pyrenophora* spp. The first approach was random amplified polymorphic DNA (RAPD) which has already been shown to be a valuable technique for obtaining specific primers for *P. teres* f. sp. *maculata* (Stevens *et al.*, 1996). As this method was only partially successful a second approach was also employed. This was based on the internal transcribed spacer (ITS) region of the nuclear ribosomal unit, which lies between the 17S and 25S rRNA genes and contains two variable non-coding spacers and the 5.8S gene (Gardes and Bruns, 1996). These ITS regions have been used elsewhere for identification of plant pathogens (Xue *et al.*, 1992; Lee *et al.*, 1993; Morton *et al.*, 1995).

MATERIALS AND METHODS

Isolates of *Pyrenophora* spp. were isolated from barley seed that was obtained from different countries, as shown in Table 18.1. The cultures were grown and DNA extracted as previously described (Stevens *et al.*, 1996). For each of the isolates listed in the table, sequencing reactions were performed using the universal ITS primers designed for this purpose (White *et al.*, 1990). An initial PCR was done using the outer ITS4 and ITS5 primers, generating a fragment approximately 600 bp in length (Stevens *et al.*, 1996). This was purified using the Qiaquick PCR product purification system (Qiagen), as described by the

Table 18.1. Sources and isolates of *Pyrenophora* spp. used in this study.

Species	Isolate	Country of origin	Original host
Pyrenophora graminea	Pg5	Denmark	barley
	Pg6	UK	barley
	Pg7	UK	barley
	Pg8	UK	barley
	Pg11	UK	barley
	Pg13	UK	barley
	Pg15	UK	barley
	Pg2376	Norway	barley
Pyrenophora teres f. sp. *maculata*	Pm2	Denmark	barley
	Pm3	Denmark	barley
	Pm4	Denmark	barley
Pyrenophora teres	Pt4	UK	barley
	Pt15	UK	barley
	Pt16	UK	barley
	Pt17	UK	barley
	Pt18	UK	barley
	Pt4035	Norway	barley
Pyrenophora avenae	Pa3828	Norway	oats
	Pa3832	Norway	oats
	Pa3833	Norway	oats
	Pa3838	Norway	oats

manufacturer, and the purified product was used as a template for sequencing. Four sequencing reactions were performed for each isolate using the universal primers ITS1–ITS4 inclusive, in conjunction with the sequencing enzyme Amplitaq® DNA polymerase FS (Perkin–Elmer), as directed in the manufacturer's protocol. The subsequent sequence analysis was done using DNASTAR software.

RESULTS AND DISCUSSION

Primers from the conserved regions of the 17S and 25S rDNA successfully amplified a single DNA product which included the ITS1 and ITS2 regions as well as the 5.8S gene. The length of the entire ITS1 and ITS2 regions and 5.8S gene from sequence data was 536–538 bp for all the isolates of the species assessed. Two species of wilt pathogen, *Verticillium albo-atrum* and *V. dahliae*, have also been shown to have ITS regions of similar size (Nazar *et al.*, 1991). This is in contrast to the ITS1 region of *Leptosphaeria maculans* which differs in size between highly and weakly virulent isolates (Xue *et al.*, 1992).

A summary of the sequence analysis (DNASTAR software, Madison, Wisconsin, USA) for the ITS 1 region of all three pathogens is shown in Table 18.2.

Table 18.2. Percentage sequence similarity between the ITS1 region of *Pyrenophora* spp. isolates using the CLUSTAL method of analysis (Higgins and Sharp, 1989).

	*Pg	Pg5	Pg8	Pg11	Pg2376	Pg15	Pm2	Pm4	Pt4	Pt15	Pt17	Pt18	*Pa
*Pg	–	97.9	98.5	97.9	100	97.9	98.5	98.5	99.5	99.5	99.5	99.5	71.3
Pg5		–	98.5	100	97.9	100	98.5	98.5	97.4	97.4	97.4	97.4	71.8
Pg8			–	98.5	98.5	98.5	100	100	99.0	99.0	99.0	99.0	71.3
Pg11				–	97.9	100	98.5	98.5	97.4	97.4	97.4	97.4	71.8
Pg2376					–	97.9	98.5	98.5	99.5	99.5	99.5	99.5	71.3
6						–	98.5	98.5	97.4	97.4	97.4	97.4	71.8
Pg15							–	100	99.0	99.0	99.0	99.0	71.3
Pm2								–	99.0	99.0	99.0	99.0	1.3
Pm4									–	100	100	100	70.8
Pt4										–	100	100	70.8
Pt15											–	100	70.8
Pt17												–	70.8
Pt18													–
*Pa													

*Sequence data used from EMBL database.
Pg, *Drechslera graminea* (EMBL accession number X78124); Pa, *Drechslera avenae* (EMBL accession number X78123).

The sequences of a *P. graminea* isolate and a *P. avenae* isolate from the EMBL database are also included (Table 18.2). Table 18.2 illustrates the high degree of taxonomic relatedness between the three *Pyrenophora* spp. pathogenic on barley as they all show above 97% similarity with each other for ITS1. It has been demonstrated that all three *Pyrenophora* spp. are freely able to hybridize interspecifically and produce fertile progeny under laboratory conditions (Smedegaard-Petersen, 1983). Further studies under natural conditions need to be carried out to determine whether the components of this pathogen complex should actually be considered as physiological races of the same biological species. Table 18.2 also shows that there are significant genetic differences between the ITS regions of *Pyrenophora* spp. pathogenic on barley and *P. avenae*, an oat pathogen. A difference in host range and ecological niche may explain the sequence divergence.

In other filamentous fungi sufficient variation was found in the ITS region to design primers for differentiation and diagnosis (Xue *et al.*, 1992; Hu *et al.*, 1993). Unfortunately, there are insufficient base pair differences between these three *Pyrenophora* species to design specific primers for interspecific differentiation. However, the genetic differences between *P. avenae* and the other *Pyrenophora* spp. should enable primers to be designed to discriminate *P. avenae* from the species pathogenic on barley. These primers could then be used to develop a quick primary screen to detect *Pyrenophora* spp. on barley seed before doing a secondary test to differentiate the three pathogens. Future studies will concentrate on analysis of the intergenic spacer region (IGS) for this pathogen complex which may be more useful for discriminating these closely related species (Appel and Gordon, 1995).

The long-term goal of this research is to develop a multiplex PCR seed health test which will be used to detect and differentiate *Pyrenophora* spp. pathogenic on barley. The test must be quantitative because it is necessary to determine the level of infection in the seed sample. This information aids a farmer to make informed management decisions such as whether seed treatment or foliar spraying is more appropriate. Quantification of PCR products has been successfully achieved for *Verticillium* spp. by using competitive PCR with an internal control template (Hu *et al.*, 1993).

The specific information required concerns the number of infected seeds in a test sample rather than the total amount of pathogen inoculum carried on the seed. This is difficult to determine using molecular techniques which would use bulk seed as it is not practical to use individual seeds. A statistical approach such as the method of maximum likelihood or the most probable number technique (Taylor, 1970) may aid the development of the most accurate method of evaluating levels of seed infection.

The presence of an internal control template is also useful for identifying false negative PCR results which may be due to the presence of substances inhibitory to PCR in the test sample. A new seed health test for barley would probably involve soaking seed to release the pathogen into a liquor, aliquots of which would be used for PCR. Inhibition of PCR using barley seed soak liquor is a known problem (Stevens *et al.*, 1996) and methods of preventing it are being investigated, e.g. dilution of liquor used for PCR, heating the liquor aliquot before PCR to denature problematical proteins and the addition of bovine serum albumin to the PCR reaction mixture.

Ultimately, we aim to automate the developed seed health test and to eliminate the need for a gel electrophoresis step. One potential method is the use of a sensitive fluorescence detection system such as Taqman (Perkin-Elmer) (Livak *et al.*, 1995). It is anticipated that with further development an improved and more cost-effective test for *Pyrenophora* spp. will become available to agriculture. This will improve the quality of seed available to farmers and reduce the spread of these pathogens.

ACKNOWLEDGEMENTS

This research was supported by a MAFF grant. We would like to thank Robert Feakes at the Department of Genetics, University of Cambridge, for his assistance with sequencing, Karen Husted and the International Mycological Institute for providing some of the *Pyrenophora* spp. isolates, and members of the Molecular Biology and Diagnostics Section, NIAB, for their technical support.

REFERENCES

Appel, D.J. and Gordon, T.F. (1995) Intraspecific variation within populations of *Fusarium oxysporum* based on RFLP analysis of the intergenic spacer region of the rDNA. *Experimental Mycology* 19, 120–128.

Blakemore, E.J.A. and Reeves, J.C. (1993) PCR used in the development of a new seed health test to identify *Erwinia stewartii*, a bacterial pathogen of maize. In: Sheppard, J.W. (ed.), *Proceedings of First ISTA–Plant Disease Committee Symposium, Ottawa, Canada*, pp. 19–22.

Cockerell, V., Rennie, W.J. and Jacks, M. (1995) Incidence and control of leaf stripe (*Pyrenophora graminea*) in Scottish barley during the period 1987–1992. *Plant Pathology* 44, 655–661.

Gardes, M. and Bruns, T.D. (1996) ITS-RFLP matching for identification of fungi. In: Clapp, J.P. (ed.), *Species Diagnostics Protocols; PCR and other Nucleic Acid Methods.* Humana Press Inc., New Jersey, pp. 177–186.

Higgins, D.G. and Sharp, P.M. (1989) Fast and sensitive multiple sequence alignments on a microcomputer. *Computer Applications in the Biosciences* 5, 151–153.

Hu, X., Nazar, R.N. and Robb, J. (1993) Quantification of *Verticillium* biomass in wilt disease development. *Physiological and Molecular Plant Pathology* 42, 23–36.

Jaccoud, D.S. and Reeves, J.C. (1993) Detection and identification of *Phomopsis* species in soya bean seeds using PCR. In: Sheppard, J.W. (ed.), *Proceedings of First ISTA–Plant Disease Committee Symposium, Ottawa, Canada*, pp. 34–43.

Jordan, V.W., Best, G.R. and Allen, E.C. (1985) Effects of *Pyrenophora teres* on dry matter production and yield components of winter barley. *Plant Pathology* 34, 200–206.

Lee, S.B., White, T.J. and Taylor, J.W. (1993) Detection of *Phytophthora* species by oligonucleotide hybridization to amplified ribosomal DNA spacers. *Phytopathology* 83, 177–181.

Livak, K., Marmaro, J. and Flood, S. (1995) Guidelines for designing Taqman fluorogenic probes for 5′ nuclease assays. *Perkin Elmer Research News*, April, 1–6.

Morton, A., Carder, J.H. and Barbara, D.J. (1995) Sequences of the internal transcribed spacers of the ribosomal RNA genes and relationships between isolates of *Verticillium albo-atrum* and *V. dahliae. Plant Pathology* 44, 183–190.

Nazar, R.N., Hu, X., Schmidt, J., Culham, D. and Robb, J. (1991) Potential use of PCR-amplified ribosomal intergenic sequences in the detection and differentiation of verticillium wilt pathogens. *Physiological and Molecular Plant Pathology* 39, 1–11.

Nicholson, P., Lees, A.K., Maurin, N., Parry, D.W. and Reazanoor, H.N. (1996) Development of a PCR assay to identify and quantify *Microdochium nivale* var. *nivale* and *Microdochium nivale* var. *majus* in wheat. *Physiological and Molecular Plant Pathology* 48, 257–271.

Rennie, W.J. and Tomlin, M.M. (1984) Barley leaf stipe. Working Sheet no. 6, 2nd edn. *ISTA Handbook on Seed Health Testing*. ISTA, Zurich.

Smedegaard-Petersen, V. (1983) Cross fertility and genetic relationship between *Pyrenophora teres* and *P. graminea*. The causes of net blotch and leaf stripe of barley. *Seed Science and Technology* 11, 673–680.

Stevens, E.A., Alderson, J., Blakemore, E.J.A. and Reeves, J.C. (1996) Development of a multiplex PCR seed health test to detect and differentiate three pathogens of barley.

In: Marshall, G. (ed.), *Diagnostics in Crop Protection*. ECPC Symposium no. 65. BCPC, University of Warwick, pp. 99–104.

Taylor, J. (1970) The quantitative estimation of the infection of bean seed with *Pseudomonas phaseolicola* (Burkh.) Dowson. *Annals of Applied Biology* 66, 29–36.

White, T.J., Bruns, T., Lee, S. and Taylor, J. (1990) Amplification and direct sequencing of fungal ribosomal RNA genes for phylogenetics. In: Innis, M.A., Gelfand, D.H., Sninsky, J.J. and White, T.J. (eds), *PCR Protocols: A Guide to Methods and Applications*. Academic Press, Totowa, New Jersey, pp. 315–322.

Xue, B., Goodwin, P.H. and Annis, S.L. (1992) Pathotype identification of *Leptosphaeria maculans* with PCR and oligonucleotide primers from ribosomal internal transcribed spacer sequences. *Physiological and Molecular Plant Pathology* 41, 179–188.

Characterization of the Seedborne Fungal Pathogen *Phomopsis phaseoli* f. sp. *meridionalis*: the Agent of Soybean Stem Canker

19

D.S. Jaccoud-Filho[1], D. Lee[2], J.C. Reeves[2] and J.T. Yorinori[3]

[1]*Agronomy Department, Ponta Grossa State University, Ponta Grossa, CP 992, Parana, Cep 84010-330, Brazil;* [2]*National Institute of Agricultural Botany, Huntingdon Road, Cambridge, CB3 OLE, UK;* [3]*National Soybean Research Centre, EMBRAPA, CP 231, Londrina, Parana, Cep 860001-970, Brazil*

INTRODUCTION

Precise characterization and detection of a seedborne pathogen is an important strategy for avoiding the dissemination of infected seed to different areas and countries. The absence of sensitive and rapid seed pathological detection methods is the primary reason for the spread of new diseases throughout the world.

In Brazil a new fungal seedborne pathogen of soybean (*Phomopis phaseoli* f. sp. *meridionalis* (teleomorph. *Diaporthe phaseolorum* f. sp. *meridionalis*)) was first detected in the 1989/90 crop season (Yorinori, 1990). During this period there was intensive exchange of genetic material between the USA and Brazil (Yorinori, 1990) and this may have been responsible for the introduction of this pathogen. Since then soybean stem canker has spread to different soybean-producing regions in Brazil and has also been reported in other South American countries (e.g. Argentina, Paraguay and Bolivia). In Brazil, this disease has caused large crop losses which are typically between 50 and 80%. However, complete crop loss can occur where a susceptible variety has been cultivated. Based upon information from the National Soybean Research Centre from Brazil the economic loss caused by *P. meridionalis* in Brazil for the 1994/95

(eds J.D. Hutchins and J.C. Reeves)

crop season alone can be estimated to be more than US$170 million (Yorinori, 1996).

Despite the low level of seed infection (less than 2%), infected seed has been reported as an important source for the dissemination of *P. meridionalis* over long distances (Yorinori, 1990). Likewise, crop debris and agricultural machinery are important for the proliferation of this pathogen in already infected production areas. Infection by soybean stem canker initially appears as small reddish-brown lesions in the stems, which develop into large elliptical lesions. The medulla becomes reddish-brown in colour and as the disease develops interveinal necrotic areas can be observed on the soybean leaves. *P. meridionalis* disturbs the translocation of nutrients in the vascular systems, causing the soybean plants to wilt and die. Artificially infected seed showed dense and appressed mycelia, light to dark-brown in colour with chlamydospore strands on the seed coat (Yorinori, 1992). Dark-brown to nearly black pycnidia can be seen within or slightly erumpent (bursting through the surface) to the seed coat which frequently open into long splits or cracks. Only alpha-type conidia are observed and these are frequently malformed (atypical) (Yorinori, 1992).

The use of traditional seed pathology ('blotter') tests have been shown to be inaccurate and insensitive in differentiating between the different *Phomopsis* spp. of the *Phomopsis*/*Diaporthe* complex which infect soybean (e.g. *Phomopsis phaseoli* f. sp. *sojae*; *Diaporthe phaseolorum* f. sp. *caulivora* and *Phomopsis longicolla*). Furthermore, the blotter test requires a 7–10 day incubation period and the analysis is often subjective. The use of random amplified polymorphic DNAs (RAPDs) (Williams *et al.*, 1990) to identify DNA fragments specific to a species/organism has aided progress in the identification and detection of a range of pathogens (e.g. see Blakemore and Reeves, 1993; Nicholson *et al.*, 1996). The advantages of using RAPD is that no previous nucleotide sequence information is required for the design of primers and polymorphisms can be demonstrated between strains and species (Fani *et al.*, 1993; Foster *et al.*, 1993). Bands identified in RAPD profiles can be sequenced and subsequently primers may be designed to amplify the DNA fragment in PCR as part of a diagnostic test (Crowhurst *et al.*, 1991; Blakemore and Reeves, 1993; Jaccoud-Filho, 1996; Jaccoud-Filho, unpublished results).

MATERIALS AND METHODS

A range of isolates of *P. meridionalis*, *P. sojae*, *P. longicolla* and *D. caulivora* were obtained from different countries (i.e. Brazil, Argentina, Paraguay, Bolivia, USA, Tanzania, Nigeria and Italy). The isolates were firstly characterized by morphological and pathological methods as previously described (Jaccoud-Filho, 1996). Subcultures of the isolates were grown on potato dextrose agar (PDA) plates and incubated at 25–27°C with a 12 h photoperiod of fluorescent light for 35 days. The mycelial characteristic was observed at 7 days and other cultural characteristics, including kind of conidia, and size of conidia, pycnidia and perithecia produced, were assessed after 35 days.

Pathogenicity tests were conducted using the toothpick method previously described by Crall (1952) and Yorinori (1991). Soybean plants (15 days old) were inoculated with a toothpick containing the *Phomopsis* mycelium and incubated for 25 days in a growth chamber (25–27°C, 15 h light/9 h dark). Symptoms of pathogenicity were evaluated by observation of the lesions at the point of inoculation in the soybean stem.

DNA used to generate RAPD and PCR profiles was obtained directly from mycelium or conidia, extracted as described by Raeder and Broda (1985). DNA oligonucleotide primers were purchased from Genosys (Cambridge). Conditions for PCR were as described by Jaccoud-Filho (1996). Random decamer primers were screened using isolates of the different *Phomopsis/Diaporthe* species. The RAPD profiles generated were useful for differentiating and characterizing isolates of *P. meridionalis* from the other components of the *Phomopsis/Diaporthe* complex (Jaccoud-Filho, 1996; Jaccoud-Filho, 1996, unpublished results).

RESULTS AND DISCUSSION

Morphological and pathological analyses were useful for the initial characterization of isolates. This served to identify unknown isolates and verify those which had been previously identified. Isolates of *P. meridionalis* characteristically had white to dark-brown mycelium (depending on age) and pycnidia with short or absent beaks which opened with a long split. Only alpha conidia were observed. Alpha conidia are normally elliptical and hyaline with an oil drop at each end. However, it has been reported by Yorinori (1992) that the alpha conidia of *P. meridionalis* are normally irregular in shape. The results obtained during this research confirm this observation in the analysis of a range of *P. meridionalis* received from Brazil and other South American countries. However, this conidial characteristic was also reported by Morgan-Jones (1984) in some isolates of *P. longicolla*. The production of the teleomorph phase *Diaporthe* was inconsistent between the *P. meridionalis* isolates.

Isolates of *P. sojae* had white to light-brown mycelium, with pycnidia opening into an apical ostiole with short or absent beaks. Both alpha and beta conidia were observed. No isolates of *P. sojae* developed the perithecial phase even when grown on PDA plates with a section of sterile soybean stem.

P. longicolla was characterized by the production of pycnidia with relatively long beaks. Only alpha conidia were observed, although the occurrence of beta conidia has been described (Hobbs *et al.*, 1985). The teleomorph phase was not observed in this group of isolates. This observation agrees with the description of this species by Hobbs *et al.* (1985). Only the perithecial stage was observed in isolates of *D. caulivora*; the anamorph stage was rarely produced.

The plasticity of the morphological characters and the large number of hosts infected by the *Phomopsis* spp. (Morgan-Jones, 1992; Rehner and Uecker, 1994) explain why the use of the morphological and pathological attributes often do not conclusively identify an isolate. The overlapping of conidial sizes between the different species makes this an unreliable character for species differentiation (Jaccoud-Filho, 1996). However, morphological and pathological characteristics are important for preliminary identification. RAPDs have proved to be more reliable for the identification of *Phomopsis* spp. One primer consistently amplifies a strong band of 900 bp from isolates of *P. meridionalis*. Four distinct clones were obtained from this band, one of which was shown to be indicative of *P. meridionalis*. Primers were designed from this fragment and a specific PCR-based test for *P. meridionalis* developed.

Sensitivity assays using purified DNA and conidial samples have shown that the test can detect around ten genomes (Jaccoud-Filho, 1996). These numbers do not take into account any inhibitory effects to the PCR from using seed soak liquor. Experiments with infected soybean seeds have demonstrated the utility of these primers for detecting of *P. meridionalis* directly from infected seed. Even though the use of PCR can be highly sensitive and accurate in detecting a specific pathogen in a seed sample it cannot determine whether the pathogen is living or dead. By using 'competitive' PCR (e.g. Hu *et al.*, 1995) it may also be possible to measure the quantity of pathogen in the seed sample but it will not be possible to determine its distribution. This problem may be overcome by using a statistical approach (e.g. most probable number). In view of the huge losses to Brazilian growers caused by *P. meridionalis*, a positive/negative test as offered by this PCR-based method may still prevent further dissemination of infected soybean seed. Further experiments are being carried out in Brazil with a range of soybean samples and different reaction conditions with the objective of optimizing the PCR seed health test.

ACKNOWLEDGEMENTS

The results of this research form part of the PhD thesis project of the first author at NIAB and University of Cambridge. This was supported by the Brazilian National Research Council (CNPq/RHAE). We would like to thank the following persons for sending isolates of *Phomopsis* spp.: Drs Jose T. Yorinori, Leo P. Ferreira (CNPSo); Drs Marcelo C. Canteri and Dauri Tessmann (UEPG); Drs Heloisa Moraes and Jose O. Menten (ESALQ); Luiz B. Nasser and Roberto T. Alves (CPAC); Drs Jose C. Machado, Maria Laene, Maria Vieira and Sonia Arias (UFLA); Celso Dornelles (CNPGC); Jose L. Gomes and Luiz A. Maffia (UFV) (all from Brazil); Mrs A. E. Eldridge (IMI, UK); Drs Antonio Ivancovich, Dora Barreto and Beatriz Perez (INTA, Argentina); Drs Dennis McGee and Anne Iles

(ISU, USA); Prof. Schmitthenner (OSU, USA); Mr David Ojo (IITA, Nigeria); Dr Rita Resca (AGRA, Italy). We would like also to thanks the members of the Molecular Biology and Diagnostic Section (NIAB) and the Plant Science Department (University of Cambridge).

REFERENCES

Blakemore, E.J.A. and Reeves, J.C. (1993) PCR used in the development of a new seed health test to identify *Erwinia stewartii*, a bacterial pathogen of maize. In: Sheppard, J.W. (ed.) *Proceedings of the First ISTA Plant Disease Committee Symposium, Ottawa, Canada*, pp. 19–22.

Crall, J.M. (1952) A toothpick tip method of inoculation. *Phytopathology* 42, 5–6.

Crowhurst, R.N., Hawthorne, B.T., Rikkerink, E.H.A. and Templeton, M.D. (1991) Differentiation of *Fusarium solani* f. sp. *cucurbitae* races 1 and 2 by random amplification of polymorphic DNA. *Current Genetics* 30, 391–396.

Fani, R., Damiani, G., Serio, C.D.I., Gallori, E., Grifoni, A. and Bazzicalupo, M. (1993) Use of random amplified polymorphic DNA (RAPD) for generating specific DNA probes for microorganisms. *Molecular Ecology* 2, 243–250.

Foster, L.M., Kozak, L.K., Loftus, M.G., Stevens, J.J. and Ross, I. (1993) The polymerase chain reaction and its application to filamentous fungi. *Mycological Research* 97, 769-781.

Hobbs, T.W., Schmitthenner, A.F. and Kuter, G.A. (1985) A new *Phomopsis* species from soybean. *Mycologia* 77, 535–544.

Hu, X., Lai, F.M., Reddy, S.N. and Ishimaru, C.A. (1995) Quantitative detection of *Clavibacter michiganensis* by competitive polymerase chain reaction. *Phytopathology* 85, 1468–1473.

Jaccoud-Filho, D.S. (1996) Identification, differentiation and detection of *Phomopsis phaseoli* f. sp. *meridionalis* in soya bean seed. PhD thesis, University of Cambridge, Cambridge, 181 pp.

Morgan-Jones, G. (1984) The *Diaporthe/Phomopsis* complex of soybeans: morphology. In: *Proceedings of the Conference Diaporthe/Phomopsis Disease Complex of Soyabean.* Agricultural Research, USDA, Washington, DC, pp. 1–7.

Morgan-Jones, G. (1992) The *Diaporthe phaseolorum* complex of soybean. *Fitopatologia Brasileira* 17, 359–367.

Nicholson, P., Lees, A.K. Maurin, N., Parry, D.W. and Reazanoor, H.N. (1996) Development of a PCR assay to identify and quantify *Microdochium nivale* var. *nivale* and *Microdochium nivale* var. *majus* in wheat. *Physiological and Molecular Plant Pathology* 48, 257–271.

Raeder, U. and Broda, P. (1985) Rapid preparation of DNA from filamentous fungi. *Letters in Applied Microbiology* 1, 17–20.

Rehner, S.A. and Uecker, F.A. (1994) Nuclear ribosomal internal transcribed spacer phylogeny and host diversity in the coelomycete *Phomopsis. Canadian Journal of Botany* 72, 1666–1674.

Williams, J.G.K., Kubelik, A.R., Livak, K.J., Rafalski, J.A. and Tingey, S.V. (1990) DNA polymorphisms amplified by abitrary primers are useful as genetic markers. *Nucleic Acids Research* 18, 6531–6535.

Yorinori, J.T. (1990) *Cancro da Haste da Soja*. Comunicado Tecnico no. 44, EMBRAPA-CNPSo, Londrina, Brazil, pp. 1–8.

Yorinori, J.T. (1991) Metodologia de produção de inóculo de *Diaporthe phaseolorum* f. sp. *meridionalis*. In: *Resumos do XXIV Congresso Brasileiro de Fitopatologia, Rio de Janeiro, Brasil.*

Yorinori, J.T. (1992) Identification of *Diaporthe phaseolorum* f. sp. *meridionalis* on soybean seeds by blotter tests. In: *Proceedings of the 23rd ISTA Congress Symposium*, Buenos Aires, Argentina, p. 107.

Yorinori, J.T. (1996) *Cancro da Haste da Soja: Epidemiologia e Controle*. Circular Tecnica no. 14, EMBRAPA-CNPSo, Londrina, Brazil.

Detection of *Tilletia indica* in Wheat and *T. barclayana* in Rice Samples and its Implication for Seed Certification

20

S.S. Chahal and P.P.S. Pannu

Department of Plant Pathology, Punjab Agricultural University, Ludhiana-141 004, India

INTRODUCTION

Tilletia indica Mitra causes Karnal bunt of wheat (*Triticum aestivum* L.) whereas *T. barclayana* (Bref.) Sacc. & Syd. causes kernel smut of rice (*Oryza sativa* L.). Both these diseases are important and are seedborne as well as soilborne, which has quite wide implications on the movement of these cereals (Warham, 1986). Karnal bunt of wheat has been reported from Pakistan, Nepal, Mexico, India, Afghanistan and Iraq (Warham, 1986; Singh *et al.*, 1989) whereas kernel smut of rice has been reported from Sierra Leone, Burma, China, India, Indonesia, Japan, Korea, Malaysia, Pakistan, The Philippines, Australia, Fiji, Greece, Mexico, USA and Trinidad. Both pathogens cause loss in yield and seedling vigour and deterioration in the quality of the produce.

Detection of these pathogens is based primarily on the presence of teliospores in or on the seeds. However, identification based on teliospore morphology is not always accurate and reliable because of morphological similarities between the teliospores of these pathogens. *T. barclayana* is sometimes also found (as a contaminant) in wheat samples (Matsumoto *et al.*, 1984; Smith *et al.*, 1996). Therefore, reliable, specific and practicable methods are needed for accurate detection and identification of these fungi. This paper deals with detection of *Tilletia* spp. in wheat and rice seed samples through various methods, namely visual and microscopic examination, washing test, seed soak method, viability test, filter and centrifuge extraction and nucleic acid based methods.

LOCATION OF THE PATHOGENS

In infected wheat and rice plants some of the ears are infected and in infected ears usually only a few grains are diseased with the seeds being partially or completely infected. The teliospores remain covered by the pericarp in wheat, whereas in rice they are covered by glumes. When the pericarp or glumes are broken, abundant teliospores are released and adhere individually or in groups to the surface of non-infected seeds.

DETECTION OF INFECTED SEEDS

Visual Inspection

Infected seeds of wheat and rice can easily be detected by examining seed samples in daylight. The range of infection varies from the small area adjoining the embryo tip to complete conversion of the seed into a black powdery mass of teliospores (Mathur and Cunfer, 1993). Sometimes masses of teliospores exude and remain adhering to the rice glumes.

Microscopic Examination

Teliospores of both pathogens generally remain adhering to the surface of non-infected, normal looking seeds. Therefore, they can be detected by observing the seeds very carefully under a stereoscopic microscope at 12.25× magnification (Begum and Mathur, 1989). Further, the contaminated seeds may be subjected to the washing test (see below) to confirm the identity of pathogen.

Washing Test

The washing test is useful for rapid detection of surface-borne teliospores of both *T. indica* and *T. barclayana* (Shetty *et al.*, 1988; Begum and Mathur, 1989). Spores are washed from seeds in water containing one or two drops of detergent. The suspension is then centrifuged, the supernatant discarded and the pellet resuspended in 5 ml of Shear's solution (Mathur and Cunfer, 1993). This suspension is then examined under a compound microscope for the presence of fungal teliospores (Begum and Mathur, 1989). For a quantitative estimation, the spore load per seed is estimated with the aid of a haemocytometer.

NaOH Seed Soak Method

Detection of diseased seeds having a low level of infection or seeds covered with dust can be difficult using the previously described methods. With such samples, bunted seeds can be detected using the NaOH soak method (Agarwal and Verma, 1983; Agarwal and Srivastava, 1985). The seeds are soaked in 2% NaOH solution for 20 h at 20°C. After this, the solution is decanted and the seeds thoroughly washed in tap water. Discoloration of infected seeds becomes clearly evident. The infected portion of the seeds appear jet black and shiny, in contrast to the pale yellow healthy seed. When black seeds are ruptured in a drop of water a stream of teliospores is released. This method can be used even when seed samples have been treated with coloured fungicides: the NaOH treatment removes the dye, clears the seed surface and increases the colour contrast between diseased and healthy seeds (Agarwal and Mathur, 1992).

Filter and Centrifuge Extraction Technique

In this method, described by Castro *et al.* (1994), a sample of seeds is washed in water containing 0.001% Tween 20. The wash water is filtered through a 60 μm diameter pore size screen and then through 12 μm pore size membrane. The first wash separates out debris and the second teliospores. The spores are then washed from the membrane and pelleted by centrifugation. The spore pellet is collected and suspended in water for examination under the light microscope for identification of the spores. As few as five spores per 100 seeds can be detected using this method (Castro *et al.*, 1994).

Further Characterization of Teliospores

In a laboratory environment both teliospore viability and isozyme characterization can be carried out. Teliospores viability can be tested by assessing germination on 2% water agar (Chahal and Mathur, 1992). Single teliospore cultures (spores or mycelium from germinated spores) can be characterized (Micales *et al.*, 1986; Bonde *et al.*, 1989). Large differences in the isozymes of *T. indica* and *T. barclayana* have shown the usefulness of this technique for distinguishing the teliospores of these two pathogens.

Nucleic Acid Based Techniques

Several attempts have been made to develop species specific polymerase chain reaction (PCR) primers for fungal plant pathogens (Hensen and French, 1993). Fungal mitochondrial DNA has been widely used as a source of molecular markers for evolution (Bruns *et al.*, 1991), taxonomy (Martin and Kistler, 1990) and genetic diversity studies (Forster and Coffey, 1993). This technique has been used to differentiate *T. indica* and *T. barclayana* (Smith *et al.*, 1996). Mitochondrial DNA has been extracted from isolates of *T. indica* and used to design (select) primers. These primers are specific for *T. indica* and not other *Tilletia* species. This approach has been used to identify germinating teliospores (Ferreira *et al.*, 1996; Smith *et al.*, 1996). These primers have the potential to be used in a practical assay with DNA extracted from teliospores contaminating wheat seed samples.

COMPARISON OF TEST METHODS

The relative efficiency of four methods, namely visual observation, microscopic examination, washing test and NaOH soak method, were compared in the laboratory. Subsamples of a seed sample, infected with Karnal bunt, were tested at the Danish Government Institute of Seed Pathology for Developing Countries (DGISP), Copenhagen, Denmark and at the Punjab Agricultural University (PAU), Ludhiana, Punjab, India. The data are presented in Table 20.1. Teliospores of *T. indica* were detected by all four methods. Generally there was agreement between the two laboratories. The NaOH sodium soak method was found to be superior to the visual observation method, both laboratories detecting teliospores more often using the former method. These methods are qualitative except for the washing test where the quantity of teliospores detected was determined to assess the spore load in the samples.

QUARANTINE SIGNIFICANCE AND IMPLICATIONS FOR SEED CERTIFICATION

Detection and accurate identification of teliospores of *Tilletia* spp. is of great quarantine importance. The causal agent of kernel smut of rice (*T. barclayana*) is a quarantine pathogen in Australia (Neergaard, 1980) whereas *T. indica*, the pathogen of wheat Karnal bunt, is a subject of strict quarantine regulations in countries free from Karnal bunt including the USA (Warham, 1986; Anon., 1991). It has often been found that *T. barclayana* contaminates wheat shipments because the wheat transport system, harvesting equipment, storage

Table 20.1. Comparison of the detection of teliospores of *Tilletia indica* with four methods from samples of an infected seed lot.

	Observations[a]	
Method	DGISP	PAU[b]
Visual observation		
Number of seeds evaluated	1498	400
Coefficient of infection (%)	4.1	4.0
Incidence (%)	10.3	9.7
NaOH		
Number of seeds evaluated	400	400
Coefficient of infection (%)	5.3	6.8
Incidence (%)	15.75	16.5
Microscopic examination		
Number of seeds evaluated	1344	400
Seed with teliospores (%)	100	100
Washing test		
Number of seeds evaluated	400	400
Number of teliospores per ml	2,210,000	2,200,000

[a]Tests conducted at DGISP, Copenhagen (Denmark) and PAU, Ludhiana (India).
[b]Mean of two sets.

facilities and processing facilities are interchangeable for wheat and rice (Smith *et al.*, 1996). In the absence of a reliable technique, it is possible to misidentify the morphologically similar teliospores of *T. indica* and *T. barclayana* which could lead to the introduction of the Karnal bunt pathogen into a disease-free area, due to the misidentification as the rice kernel smut pathogen. Isozyme analysis method, using protein extracted from germinated teliospores, is considered of less practical value for routine identification of *Tilletia* species because it requires considerable experience. Therefore, the PCR-based method may provide a more reliable and correct identification of various species of *Tilletia* which is a primary requirement in quarantine regulations.

The Central Seed Certification Board in India has fixed standards for these pathogens. For foundation seed the tolerance is of 0.1% infection for both pathogens, whereas it is 0.25% for *T. indica* and 0.5% of *T. barclayana* for certified seed (Tunwar and Singh, 1988). Agricultural Canada has maintained a zero tolerance limit for *T. indica* (Martin, 1986). In view of the strict measures fixed by many countries, it is important that the detection technique should be both sensitive and accurate.

REFERENCES

Agarwal, V.K. and Mathur, S.B. (1992) Detection of *Tilletia indica* (Karnal bunt) wheat seed samples treated with fungicides. *FAO Plant Protection Bulletin* 40, 148.

Agarwal, V.K. and Srivastava, K.D. (1985) NaOH seed soak method for routine examination of rice seed lots for rice bunt. *Seed Research* 13, 159–161.

Agarwal, V.K. and Verma, H.S. (1983) A simple technique for detection of Karnal bunt infection in wheat seed samples. *Seed Research* 11, 110–112.

Anonymous (1991) Quarantine procedure no. 37. *Tilletia indica. EPPO Bulletin* 21, 265–266.

Begum, S. and Mathur, S.B. (1989) Karnal bunt and loose smut in wheat seed lots of Pakistan. *FAO Plant Protection Bulletin* 37, 165–173.

Bonde, M.R., Peterson, G.L. and Matsumoto, T.T. (1989) The use of isozymes to identify teliospores of *Tilletia indica. Phytopathology* 79, 596–599.

Bruns, T.D., White, T.J. and Taylor, J.W. (1991) Fungal molecular systematics. *Annual Review of Ecology and Systemics* 22, 525–564.

Castro, C., Schaad, N.W. and Bonde, M.R. (1994) A technique for extracting *Tilletia indica* teliospores from contaminated wheat seeds. *Seed Science and Technology* 22, 91–98.

Chahal, S.S. and Mathur, S.B. (1992) Germination of deep frozen *Tilletia indica* and *Tilletia barclayana* teliospores. *FAO Plant Protection Bulletin* 40, 31–35.

Ferreira, M.A.S.V., Tooley, P.W., Hatzilloukas, E., Castro, C. and Schaad, N.W. (1996) Isolation of species-specific mitochondrial DNA sequence for identification of *Tilletia indica*, Karnal bunt of wheat fungus. *Applied and Environmental Microbiology* 62, 87–93.

Forster, H. and Coffey, M.D. (1993) Molecular taxonomy of *Phytophthora megasperma* based on mitochondrial and nuclear DNA polymorphisms. *Mycological Research* 97, 1101–1112.

Hensen, J.M. and French, R. (1993) The polymerase chain reaction and plant disease diagnosis. *Annual Review of Phytopathology* 31, 81–109.

Martin, F.N. and Kistler, H.C. (1990) Species-specific banding pattern of restriction endonuclease-digested mitochondrial DNA from the genus *Pythium*. *Experimental Mycology* 14, 32–46.

Martin, P.M. (1986) Quarantine of Karnal bunt in Canada. In: *Proceedings of the Fifth Biennial Smut Workers Workshop, April 28–30, 1986, Ciudad, Obregon, Sonora, Mexico.* CIMMYT, p. 29.

Mathur, S.B. and Cunfer, B.M. (1993) *Seed-borne Diseases and Seed Health Testing of Wheat*. Danish Government Institute of Seed Pathology for Developing Countries. Hellerup, Denmark, 168 pp.

Matsumoto, T.T., Boratynski, T.N., Showers, D.W., Higuera, D. and Luscher, D. (1984) Karnal bunt of wheat. *California Plant Pest and Disease Reporter* 3, 45–47.

Micales, J.A., Bonde, M.R. and Peterson, G.L. (1986) The use of isozyme analysis in fungal taxonomy and genetics. *Mycotaxon* 27, 405–449.

Neergard, P. (1980) A review on quarantine for seed. In: *National Academy of Sciences, India*, Golden Jubilee Commemoration Volume, pp. 495–530.

Shetty, S.A., Aruna, K. and Shetty, H.S. (1988) Investigations on kernel smut of paddy. *Plant Disease Research* 4, 172–176.

Singh, D.V., Agarwal, R., Shreshtha, J.K., Thappa, B.R. and Dublin, H.J. (1989) First Report of *Neovossia indica* on wheat in Nepal. *Plant Disease* 73, 393–399.

Smith, O.P., Peterson, G.L., Beck, R.J., Schaad, N.W. and Bonde, M.R. (1996) Development of a PCR-based method for identification of *Tilletia indica*, causal agent of Karnal bunt of wheat. *Phytopathology* 86, 115–122.

Tunwar, N.S. and Singh, S.V. (1988) *Indian Minimum Seed Certification Standards*. The Seed Certification Board, Department of Agriculture and Co-operation, Ministry of Agriculture, Government of India, New Delhi, 24 pp.

Warham, E.J. (1986) Karnal bunt disease of wheat: a literature review. *Tropical Pest Management* 32, 229–242.

21 BIO-PCR: a Highly Sensitive Technique for Detecting Seedborne Fungi and Bacteria

N.W. Schaad, M.R. Bonde and E. Hatziloukas

ARS-USDA, Foreign Disease–Weed Science, Research Unit, Fort Detrick, MD 21702-5023, USA

INTRODUCTION

The first prerequisite to the control of any disease is detection and proper identification of the pathogen. Because many seedborne pathogens are present in very low numbers in contaminated or infected seeds, detection is often difficult (Schaad, 1982; Gabrielson, 1988; Stace-Smith and Hamilton, 1988; Saettler *et al.*, 1989). A sensitive detection technique becomes even more important when seeds are treated with chemicals (Clayton, 1931), or a combination of hot water and chemical (Schaad *et al.*, 1980; Strangberg and White, 1989; Fatmi *et al.*, 1991) to eradicate the pathogen. Such treatments normally work well for infested seeds but not always for infected seeds and thus can be considered successful only if a reliable test is used to assay the treated seed for viable cells of the target pathogen. Another problem is the chemical residue left on the seeds after treatment. The most reliable method for assaying treated seed is the direct plating of disinfested seeds on to an agar medium in a 25 cm diameter Petri plate using a specially designed vacuum planter (Schaad and Kendrick, 1975; Schaad, 1989). Common sense tells us that molecular tests such as serological and classical PCR techniques that detect dead cells cannot be used for treated seeds.

TRADITIONAL METHODS

Classical seed health assays for fungi involve plating seeds directly on to moistened filter paper and observing characteristic mycelium and spores under a microscope. For bacteria, most classical assays involve soaking seeds in liquid for varying times and plating dilutions of liquid samples on to semi-selective agar media (Schaad, 1982; Saettler *et al.*, 1989). Colonies similar to the target pathogen are removed, cloned and subjected to pathogenicity tests to confirm their presumptive identity. Such assays have proven to be of limited value because accurate identification of suspect fungi or bacteria is often difficult and time consuming. Furthermore they normally work well only with seed samples which contain proportionally more target pathogens than saprophytic organisms. An additional problem with semi-selective media is the occurrence of saprophytic bacteria which often produce antibiotics or staling products which act synergistically with antibiotics in the medium and inhibit the target pathogen. Control cultures of the pathogen must be over-sprayed with the seed wash to determine if such organisms exist (Randhawa and Schaad, 1984). Serological assays such as enzyme-linked immunosorbent assay (ELISA) (Van Vuurde *et al.*, 1983) and immunofluorescence (Coleno, 1968; Schaad, 1978) have been developed and have been quite useful for rapid testing. However, false positives and false negatives are a continual problem with such tests. Antisera can be too specific and sometimes do not react with all strains or are not specific enough and cross-react with other organisms. False negatives can result from the presence of dead cells and use of highly diluted samples. Serological assays are normally much less sensitive than agar plating assays and have the major disadvantage of not resulting in viable cultures for confirming identification by pathogenicity tests.

POLYMERASE CHAIN REACTION

One of the most exciting technical developments in the field of biology during the last 50 years was the invention in 1987 of the polymerase chain reaction (PCR) by Kerry Mullis (Mullis, 1987; Mullis and Faloona, 1987). This resulted in a Nobel prize being awarded to Mullis and it has revolutionized forensic analysis and disease diagnosis. By amplifying the DNA of an organism several million times it is possible to detect the DNA from a single cell. It is even possible to detect free DNA. PCR has enabled the development of highly sensitive and accurate methods for detection of seedborne fungi (Smith *et al.*, 1996) and bacteria (Rasmussen and Wulf, 1991; Prosen *et al.*, 1993). The key to the development of PCR was the discovery of a thermostable DNA polymerase, *Taq* polymerase, in combination with the development of a computer controlled thermocycler by Perkin–Elmer. This allowed continuous amplification by accurate cycling between heating and cooling. The technique is now used routinely in all molecular laboratories for cloning, sequencing and other molecular applications. More papers have been published on PCR during the past 12 years than any other single molecular biological technique. One important requirement for PCR is the knowledge of two 10–20 base nucleotide sequences for synthesis of primers. The easiest way to obtain primers is to purchase commercially available random primers. No research is needed and only detailed testing of a large number of primers is required in order to find a pair that is specific to DNA of the target organism. However, the most reliable method of designing specific primers is to sequence a gene fragment of a character of the target organism which is known to be unique to it.

This was the approach used for designing primers specific to *Pseudomonas syringae* pv. *phaseolicola* (PSP), the causal agent of halo blight of beans (Prosen *et al.*, 1993). Knowing that this organism produced phaseolotoxin, a unique toxin required for disease development (Mitchell, 1978; Peet *et al.*, 1986), primers were designed (Prosen *et al.*, 1993) to amplify a fragment within the phaseolotoxin biosynthetic gene cluster (Peet *et al.*, 1986). Results showed that the primers were highly specific (Prosen *et al.*, 1993). By using a 'hot-start' PCR and cells rather than DNA followed by a second round of nested primers, up to five cells of the pathogen could be detected per 10 μl sample of seed washing (Prosen *et al.*, 1993). Because the phaseolotoxin gene cluster is present as a single copy, nested PCR is required to obtain an acceptable sensitivity.

Like serology, PCR is very rapid and costs much less than classical methods requiring pathogenicity tests. However, there are many problems with PCR when using natural samples. The method lacks sensitivity simply due to the use of such a small sample size (10–20 μl). Perhaps the greatest problem is the presence of PCR inhibitors in plant and animal samples that can cause false negatives. With classical PCR bean seed washes spiked with small numbers of PSP are often found to test negative, apparently due to the presence of PCR inhibitors in some of the seed washes. Reliable detection of 100 viable cells per millilitre of seed wash is not always possible. Of great concern to seed producers is the possibility of a false positive due to the presence of dead cells or free DNA from lysed cells. Since neither causes infection, a positive result is of no consequence epidemiologically.

BIO-PCR

To overcome these problems a highly sensitive PCR technique named BIO-PCR was developed (Schaad *et al.*, 1995). The BIO-PCR seed assay is quite simple. To detect PSP in bean seed, 1 kg of bean seed is soaked overnight in 1.5 l of 0.01% Tween 20 at 2–3°C. Aliquots (0.1 ml) of the seed extract are plated on to each of five plates of King *et al.* medium B and MSP agars, as described by Mohan and Schaad (1987). After 45–48 h at 22°C, each plate is washed three times with 1 ml of water and the pooled sample stored at –20°C for direct 'hot-start' PCR using primers P 5.1 and P 3.1. The 0.5 kb amplification product from a single colony is easily detected by gel electrophoresis and ethidium bromide staining. As few as 20 cells per ml of original seed washing can be detected (Table 21.1).

Advantages of BIO-PCR over classical PCR include: (i) elimination of false positives due to dead cells; (ii) elimination of false negatives due to the presence of PCR inhibitors in the seed extract; and (iii) a 100-fold increase in sensitivity.

For fungi, spores can be extracted from seed washings, placed on to agar media, and DNA extracted from the resulting mycelium and secondary spores. Smith *et al.* (1996) using primers which amplify mitochondrial DNA of *Tilletia indica* have detected as few as five teliospores per 50 g of wheat seed (Table 21.2).

Disadvantages of BIO-PCR include: (i) failure of physiologically weakened or injured cells to grow on agar media, possibly resulting in false negatives, and (ii) possible failure to relate to field disease due to its high sensitivity. We have not determined the latter but we have done a blind test by assaying 52

Table 21.1. Detection by BIO-PCR of *Pseudomonas syringae* pv. *phaseolicola* (PSP) added to bean seed extract.

Expected no. of PSP per plate[a]	No. of cfu observed per plate KB[b]		BIO-PCR 0.45 kb band[c]
	Saprophytes	PSP	
1–2	97, 108, 118, 122	2, 2, 2, 2	+, +, +, –
< 1	85, 105, 106, 115	0, 0, 0, 0	+, +, –, –
Negative control	88, 124	0, 0	–, –

[a]Cells of PSP added to seed washing and to water to determine expected number.
[b]Seeds washed and 100 μl assayed on eight plates of KB agar, as described by Mohan and Schaad (1987).
[c]Four plates washed after 48 h and three samples of 35 μl of pooled washing used for direct 'hot-start' PCR.

Table 21.2. Sensitivity of BIO-PCR in detecting *Tilletia indica* in washings of wheat seed contaminated with a known number of teliospores.

No. of spores added per sample[a]	Replicate[b]					% of samples positive
	1	2	3	4	5	
0	0	0	0	0	0	0
1	1	0	1	0	0	40
2	1	0	2	0	0	40
5	3	3	4	6	4	100
10	10	6	6	6	5	100

[a]Known numbers of teliospores were added to washings of 50 g of wheat seed.
[b]The teliospores were extracted by sieving and germinated on agar, and DNA was extracted from the resulting mycelium, as described by Smith *et al.* (1996)

previously assayed commercial seed samples obtained from Idaho and North Dakota Departments of Agriculture by both MSP agar plating and BIO-PCR. The results showed that the BIO-PCR technique compares very favourably with the classical tests (unpublished data). Commercial seed companies should have no fear that a much larger number of seed lots will be rejected by state regulatory agencies if the more sensitive BIO-PCR assay is used.

REFERENCES

Clayton, E.E. (1931) *Vegetable Seed Treatment with Special Reference to the Use of Hot Water and Organic Mercurials.* Technical Bulletin no. 183, New York Agricultural Experimental Station, Geneva.

Coleno, A. (1968) Utilisation de la technique d'immunofluorescence pour le depistage de *Pseudomonas phaseolicola* (Burkh.) Dowson dans les lots de semences contaminés. C.R. Seances Acad. Agric. Fr. 54, 1016–1020.

Fatmi, M., Schaad, N.W. and Bolkan, H.A. (1991) Seed treatments for eradicating *Clavibacter michiganensis* subsp. *michiganensis* from naturally infected tomato seeds. *Plant Disease* 75, 383–385.

Gabrielson, R.L. (1988) Fungi. *Phytopathology* 78, 868–872.

Mitchell, R.E. (1978) Halo blight of beans: toxin production by several *Pseudomonas phaseolicola* isolates. *Physiological Plant Pathology* 13, 37–49.

Mohan, S.K. and Schaad, N.W. (1987) An improved agar plating assay for detecting *Pseudomonas syringae* pv. *syringae* and *P. syringae* pv. *phaseolicola* in contaminated bean seed. *Phytopathology* 77, 1390–1395.

Mullis, K.B. (1987) US Patent no. 4,683,202 (July 8, 1987).

Mullis, K.B. and Faloona, F.A. (1987) Specific synthesis of DNA in vitro via a polymerase-catalyzed chain reaction. *Methods in Enzymology* 155, 335–350.

Peet, R.C., Lindgren, P.B., Willis, D.K. and Panopoulos, N.J. (1986) Identification and cloning of genes involved in phaseolotoxin production by *Pseudomonas syringae* pv. *phaseolicola. Journal of Bacteriology* 166, 1096–1105.

Prosen, D., Hatziloukas, E., Schaad, N.W. and Panopoulos, N.J. (1993) Specific detection of *Pseudomonas syringae* pv. *phaseolicola* DNA in bean seed by polymerase chain reaction-based amplification of a phaseolotoxin gene region. *Phytopathology* 83, 965–970.

Randhawa, P. and Schaad, N.W. (1984) Selective isolation of *Xanthomonas campestris* pv. *campestris* from crucifer seeds. *Phytopathology* 74, 268–272.

Rasmussen, O.F. and Wulf, B.S. (1991) Detection of *P. s. pisi* using PCR.. In: Durbin, R.D., Surico, G. and Mugnai, L. (eds), *Proceedings of International Working Group on* Pseudomonas syringae *pathovars, 4th Stamperia Granducale, Florence, Italy*, pp. 369–376.

Saettler, W., Schaad, N.W. and Roth, D.A. (eds) (1989) *Detection of Bacteria in Seed and other Planting Material.* The American Phytopathological Society, St Paul, Minnesota.

Schaad, N.W. (1978) Use of direct and indirect immunofluorescence tests for identification of *Xanthomonas campestris. Phytopathology* 68, 249–252.

Schaad, N.W. (1982) Detection of seedborne bacterial plant pathogens. *Plant Disease* 66, 885–890.

Schaad, N.W. (1989) Detection of *Xanthomonas campestris* pv. *campestris* in crucifers. In: Saettler, W., Schaad, N.W. and Roth, D.A. (eds): *Detection of Bacteria in Seed and Other Planting Materials*. The American Phytopathological Society, St Paul, Minnesota, pp. 68–75.

Schaad, N.W. and Kendrick, R. (1975) A qualitative method of detecting *Xanthomonas campestris* in crucifer seed. *Phytopathology* 65, 1034–1036.

Schaad, N.W., Gabrielson, R.L. and Malanax, M.W. (1980) Hot acidified cupric acetate soaks for eradication of *Xanthomonas campestris* from crucifer seeds. *Applied and Environmental Microbiology* 39, 803–807.

Schaad, N.W., Cheong, S.S., Tamaki, S., Hatziloukas, E. and Panopoulos, N.J. (1995) A combined biological and enzymatic amplification (BIO-PCR) technique to detect *Pseudomonas syringae* pv. *phaseolicola* in bean seed extracts. *Phytopathology* 85, 243–248.

Smith, O.P., Peterson, G.L., Beck, R.J., Schaad, N.W. and Bonde, M.R. (1996) Development of a PCR-based method for identification of *Tilletia indica*, causal agent of Karnal bunt of wheat. *Phytopathology* 86, 115–122.

Stace-Smith, R. and Hamilton, R.I. (1988) Viruses. *Phytopathology* 78, 875–880.

Strangberg, J.O. and White, J.M. (1989) Response of carrot seeds to heat treatments. *Journal of the American Society for Horticultural Science* 1145, 766–769.

Van Vuurde, J.W.L., van den Bovenkamp, G.W. and Birnbaum, Y. (1983) Immunofluorescence microscopy and enzyme-linked immunosorbent assay as potential routine tests for the detection of *Pseudomonas syringae* pv. *phaseolicola* and *Xanthomonas campestris* pv. *phaseoli* in bean seed. *Seed Science and Technology* 11, 547–559.

22

Immunofluorescence Colony-staining as a Tool for Sample Indexing and for the Determination of Field Thresholds for Pathogenic Seedborne Bacteria

J.W.L. Van Vuurde

DLO Research Institute for Plant Protection (IPO-DLO), Binnenhaven 12, PO Box 9060, 6700 GW Wageningen, The Netherlands

INTRODUCTION

Initiatives have recently been taken by EPPO and FAO to restrict the lists of quarantine pests to those of actual scientific and economic importance (Van Halteren, 1995). For quarantine pests, assays can be qualitative as the tolerance level is zero. The main concern will be the reliability of the test at the chosen detection level. The tolerance level should be based on the risk of economic crop losses in the region of final production. By correlating quantitative results of testing seed lots for a seedborne pest with observations on disease incidence in crops grown from the seed, threshold levels can be established for the major growing areas.

More realistic quarantine lists and the establishment of realistic thresholds for quality pests will be important to promote seed trade within the European Community and worldwide. This paper describes immunofluorescence colony-staining (IFC) as a tool for both quantitative routine indexing of samples for pathogenic seedborne bacteria and for generating laboratory test results for field threshold studies.

ASSAY CHARACTERISTICS RELATED TO FIELD INCIDENCE

To correlate laboratory test results with field disease observations, the laboratory assay should meet three requirements: it should (i) be sufficiently sensitive;

 Seed Health Testing (eds J.D. Hutchins and J.C. Reeves)

(ii) have the possibility of confirming positive results; and (iii) be able to quantify the percentage of seeds contaminated with the viable target. Estimation of the percentage of infected seeds can be based on individual seed testing or on repeated subsampling and MPN statistics (Taylor, 1970). It is important to (semi)quantify the degree of contamination of the infected seeds to understand the rate of seed transmission in the field (Taylor *et al.*, 1979; Van Vuurde *et al.*, 1991).

COMPARISON OF METHODS

Of the three methods used for detecting bacteria, namely isolation, serology and nucleic acid probes, the traditional isolation technique on agar plates matches the above-mentioned criteria most closely. The advantages of isolation assays are that only the culturable bacteria are detected and that pure cultures from suspected colonies can be confirmed and tested for pathogenicity. The major disadvantage of this technique is that detection and quantification of seedborne pathogenic bacteria in a sample are affected by the interference of non-target bacteria. Semiselective media may reduce the interference of the background saprophytes, but can also strongly affect the recovery of the various wild-types of the target bacterium (Chun and Alvarez, 1983).

The detection level of the serological techniques mostly used for seed testing, i.e. enzyme-linked immunosorbent assay (ELISA) and immunofluorescence cell staining, are generally not affected by background saprophyte interference. However, ELISA lacks sensitivity and neither test differentiates between culturable and non-culturable cells. The polymerase chain reaction (PCR) may have the potential for detecting culturable target cells when applied in combination with enrichment as in BIO-PCR (Schaad *et al.*, 1995; Chapter 21, this volume). The target may still be detectable by PCR even after overgrowth by saprophytes. However, PCR is non-quantitative and relatively sensitive to interference. Comparative tests in the field of medicine have shown an unacceptable test performance (Quint *et al.*, 1995).

IFC was designed for sensitive and quantitative detection of culturable target bacteria (Van Vuurde, 1987). It is based on a combination of isolation by pour plating (see below) with serology for target colony recognition. Confirmation of IFC-positive colonies is possible by taking cells directly from these colonies for pure culture identification or for PCR (Van der Wolf *et al.*, 1995).

PRINCIPLE OF IFC

Agar-mixed sample plating (pour plating) has the advantage over surface plating of a strongly reduced saprophytic interference. A 1 cm^2 agar layer with a thickness of about 2 mm can host about 10,000 saprophyte colonies and still allow high recoveries of the target for most target bacteria with an average colony growth rate (Van Vuurde and Roozen, 1990; Jones *et al.*, 1994). Pour plates are usually incubated for 1–2 days to obtain small colonies in the medium. Wells (16 mm diameter) of 24-well tissue culture plates are used for routine application on a large number of samples. For recognition of target colonies among non target colonies, fluorochrome-labelled antibodies proved more suitable than those labelled with enzyme or gold beads (Van Vuurde, 1990b). General media with reduced nutrient concentrations or selective media are recommended for pour plating. The IFC format described below is largely based on the procedure for non-dried agar plates (Van Vuurde and Van der Wolf, 1995).

IFC ROUTINE PROCEDURE USING MULTIWELL PLATES

Previous procedures described for IFC recommended the use of dried agar preparations with the advantage of faster staining and washing and lower agar background fluorescence (Van Vuurde, 1990a). The dried agar plates sometimes showed poor staining areas, probably due to a reduced permeability of the agar. The following procedure is based on the use of non-dried (fresh) agar plates which are more suitable for routine application of IFC and for sampling from IFC-positive colonies for confirmation.

1. Add up to 100 μl of a test sample to a 16 mm diameter well of a tissue culture plate. Add 300 μl of agar medium (1.5% agar) at 45°C to each well. Swirl the plate gently to ensure a homogeneous distribution of the bacteria in the agar.
2. Incubate the plate at the optimum temperature for the target bacterium. Observe the colony growth and recovery rate of the target in the positive control (for most target bacteria incubation at, e.g. 25°C, can be stopped after 1–2 days).
3. Add 300 μl of FITC-conjugated antiserum, diluted to a suitable working titre in 0.01 M phosphate-buffered saline with 0.1% Tween 20 (PBST), to each well and incubate while very gently shaking at room temperature for about 18 h.
4. Gently invert the plate to remove the non-bound conjugate from the wells. Wash for 1 min with 1 ml PBST per well, followed by two or more washings until the background fluorescence is sufficiently reduced. Remove the PBST from the well after the last washing.
5. Observe the wells directly under a fluorescence microscope with incident blue light and, e.g. 4× objective and 6.3× eyepiece magnification. Look for colonies with green fluorescence.
6. If necessary, verify the identity of selected IFC-positive colonies by puncturing the colony under the microscope with a fine capillary or platinum needle followed by isolation and characterization or PCR.

Recent experiments showed that the use of a thinner agar layer (e.g. 200 μl per well) and of one unit agarase per millilitre diluted conjugate allowed a stronger conjugate dilution for staining. This resulted in further improvement of the fluorescent colony to background contrast (Moore and Van Vuurde, unpublished data). Further details on the procedure, other formats for IFC, antibody conjugation, PCR and optical equipment are presented elsewhere (Van Vuurde, 1990a; Van Vuurde and Van der Wolf, 1995).

SENSITIVITY OF IFC

The use of 16 mm diameter wells of tissue culture plates limits the amount of undiluted test sample to 100 µl per well. Theoretically the detection level will be 10 cfu per millilitre of test sample if 100% recovery can be obtained. The detection level for the 16 mm diameter wells using 10–100 µl undiluted extract was determined for soft rot *Erwinia* spp. at about 10^2 cfu ml^{-1} (Van Vuurde and Roozen, 1990). Detection levels between 10 and 100 target cells ml^{-1} in undiluted soil sample extracts with a high saprophyte background were reported for *Pseudomonas* sp. (Leeman *et al.*, 1991). The recovery of slow growing organisms like *Clavibacter michiganensis* subsp. *sepedonicus* is still strongly reduced by the competition of much faster growing saprophyte colonies (Roozen and Van Vuurde, 1991). Increased sensitivity can be obtained by increasing the plate diameter and sample size.

SPECIFICITY OF TEST SYSTEMS AND OF IFC

The specificity of a test system will depend on the specificity of the probe (antibody or DNA) and of the assay format. Extracts from seed samples usually vary strongly in types and numbers of indigenous bacterial species. Increasing the sensitivity of a detection system will in general decrease the specificity as more types of saprophytic bacteria will be present in populations above that detection level. Application of more specific probes is often limited by the biological variation between wild-type populations of the bacterial pathogen.

In practice, reliable positive results with antibody- or DNA probe-based assays can be obtained by using a suitable probe to detect all isolates of the target pathogen. As unique target specificity is rare, positive results should be checked for possible false-positive reactions. Confirmation should preferably be done directly on the positive test reaction to avoid discrepancy between results of the screening and the confirmation assay. For IFC this is done for the IFC-positive colony. Two protocols were developed to confirm positive IFC results directly for cells from an IFC-positive colony (Fig. 22.1).

Reisolation for Pure Culture Characterization

As the fluorescent antibodies only mark the colony without affecting the viability of the cells inside, culturable cells can be sampled from positive colonies with

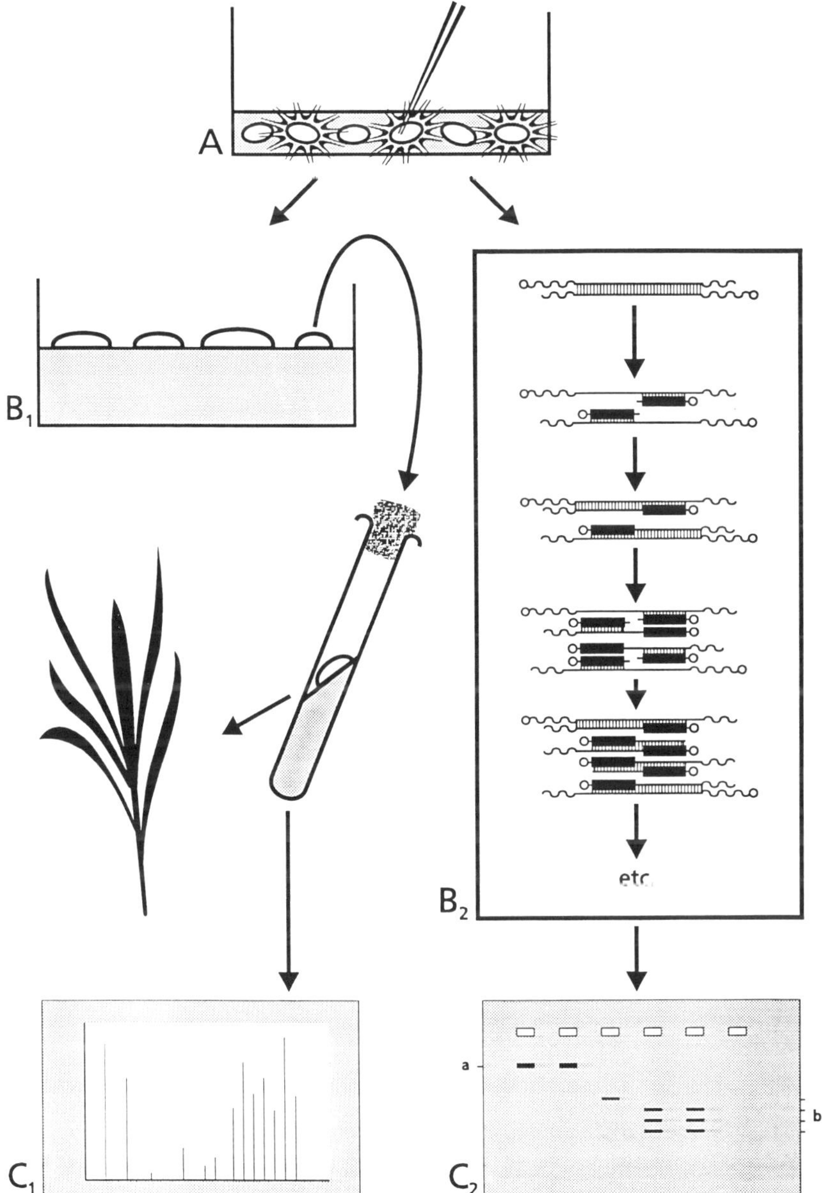

Fig. 22.1. Scheme of confirmation of IFC-positive colonies by isolation of pure cultures or by PCR. (Reproduced from *IPO-DLO Annual Report* 1992). (A) Fluorescent colonies in pour plates. Verification by (B_1) isolation from IFC-positive colonies to obtain a pure culture for e.g. biochemical and/or pathogenicity testing, or (C_1) characterization by fatty acid profiling, or (B_2) to perform PCR, prior to (C_2) characterization by gel electrophoresis.

a fine platinum or capillary needle. After pure cultures have been obtained, bacteria can be identified by reference procedures and/or can be tested for traits such as pathogenicity. Isolation from IFC-positive colonies also provides an efficient tool for selective isolation of cross-reacting bacteria. Colonies of cross-reacting bacteria are often recognizable as such by a different colony morphology and/or fluorescence staining. The efficiency of isolation is dependent upon the density of the saprophytic background and the survival of the bacterium in colonies in the test system.

Sampling from Fluorescent Colonies for Testing in PCR

This approach showed a better potential for routine confirmation of IFC-positive colonies than isolation, but lacks the possibility to obtain pure cultures for further research (Van der Wolf *et al.*, 1995). Isolated cross-reacting bacteria can be used to improve specificity if needed. Specificity can be improved by selecting monoclonal or polyclonal antisera which do not react with the cross-reacting isolates, by absorption of the antiserum with the cross-reacting strain or by making the medium inhibitory for growth of these isolates. *Bacillus cereus* var. *mycoides*, causing autofluorescence and disturbing mycoid growth in IFC for *Erwinia* spp. in seed potatoes, was effectively suppressed by the addition of 1 μg of rifampicin per litre of medium (P. Kastelein, personal communication).

SPECIAL FORMATS OF IFC

The following formats of IFC may be of interest for research on seedborne bacteria (Van Vuurde, 1990a; Van Vuurde and Van der Wolf, 1995).

- *In situ* and *in planta* IFC staining techniques have been developed to detect trace infections in agar-embedded plant parts and can be helpful for seed transmission studies (Van Vuurde *et al.*, 1994).
- The micro-IFC technique can be used separately to quantify culturable cells and non-culturable cells in samples.
- Two different seedborne pathogens in the same preparation can be detected simultaneously with a double fluorescent staining procedure.
- IFC can be combined with a previous immunomagnetic concentration of target bacteria to lower the detection level and reduce saprophytic interference (Jones and Van Vuurde, 1996).

APPLICATIONS OF IFC FOR SEEDBORNE BACTERIA

So far IFC has been developed mainly for detecting target bacteria in sample materials other than seeds (Van Vuurde *et al.*, 1992, 1995). Detection levels of *Xanthomonas oryzae* pv. *oryzae* in rice seed extracts of 10–100 cfu ml^{-1} in IFC and of 3500 cfu ml^{-1} in PCR have been reported (see Chapter 23). At IPO-DLO, IFC has only been used for detecting seedborne pathogenic bacteria in preliminary experiments, mainly during workshops and training courses; however, the technique shows good prospects for further development into a routine test for quantitative determination of culturable target bacteria in seed samples.

CONCLUSIONS

IFC provides a sensitive and reliable tool both for routine indexing of seed lots for pathogenic bacteria as for quantitative determination of culturable target bacteria for field threshold studies. Sensitivity and specificity are improved compared with traditional isolation and serological assays. Direct sampling from IFC-positive colonies is used for confirmation by isolation or PCR. Selective media may be replaced by one standard medium for a range of different seedborne bacteria in IFC as a result of the strongly reduced microbial interference between colonies in the agar.

ACKNOWLEDGEMENT

I thank the members of the IPO-DLO bacteriology group for their support, and L. Bos, H. Huttinga, P. Kastelein and J.M. Van der Wolf for their comments on the manuscript.

REFERENCES

Chun, W.C.C. and Alvarez, A.M. (1983) A starch–methionine medium for isolation of *Xanthomonas campestris* pv. *campestris* from plant debris in soil. *Plant Disease* 67, 632–635.

Jones, D.A.C., Hyman, L.J., Tumeseit, M., Smith, P. and Pérombelon, M.C.M. (1994) Blackleg potential of potato seed: determination of tuber contamination by *Erwinia carotovora* subsp. *atroseptica* by immunofluorescence colony staining and stock and tuber sampling. *Annals of Applied Biology* 124, 557–568.

Jones, J.B. and Van Vuurde, J.W.L. (1996) Immunomagnetic isolation of *Xanthomonas campestris* pv. *pelargonii*. *Journal of Applied Bacteriology* 81, 78–82.

Leeman, M., Raaijmakers, J.M., Bakker, P.A.H.M. and Schippers, B. (1991) Immunofluorescence colony staining for monitoring pseudomonads introduced in soil. In: Beemster, A.B.R., Bollen, B.J., Gerlagh, M., Ruissen, M.A., Schippers, B. and Tempel, A. (eds) *Biotic Interactions and Soil-borne Diseases*. Elsevier, Amsterdam pp. 374–380.

Quint, W.G.V., Heijtink, R.A., Schirm, J., Gerlich, W.H. and Niesters, H.G.M. (1995) Reliability of methods for hepatitis B virus DNA detection. *Journal of Clinical Microbiology* 33, 225–228.

Roozen, N.J.M. and Van Vuurde, J.W.L. (1991) Development of a semi-selective medium and an immunofluorescence colony-staining procedure for the detection of *Clavibacter michiganensis* subsp. *sepedonicus* in cattle manure slurry. *Netherlands Journal of Plant Pathology* 97, 321–324.

Schaad, N.W., Cheong, S.S., Tamaki, S., Hatziloukas, E. and Panopoulos, N.J. (1995) A combined biological and enzymatic amplification (BIO-PCR) technique to detect *Pseudomonas syringae* pv. *phaseolicola* in bean seed extract. *Phytopathology* 85, 243–248.

Taylor, J.D. (1970) The quantitative estimation of the infection of bean seed with *Pseudomonas phaseolicola* (Burkh.) Dowson. *Annals of Applied Biology* 66, 29–36.

Taylor, J.D., Dudley, C.L. and Presly, L. (1979) Studies of halo blight seed infection and disease transmission in dwarf beans. *Annals of Applied Biology* 93, 267–277.

Van der Wolf, J.M., Van Beckhoven, J.R.C.M., De Vries, P.M., Raaijmakers, J.M., Bakker, P.A.H.M., Bertheau, Y. and Van Vuurde, J.W.L. (1995) Polymerase chain reaction for verification of fluorescent colonies in immunofluorescence colony-staining. *Applied and Environmental Microbiology* 79, 569–577.

Van Halteren, P. (1995) A diagnostic network for the EPPO-region. *Bulletin OEPP/EPPO Bulletin* 25, 1–4.

Van Vuurde, J.W.L. (1987) New approach in detecting phytopathogenic bacteria by combined immunoisolation and immunoidentification assays. *EPPO Bulletin* 17, 139–148.

Van Vuurde, J.W.L. (1990a) Immunofluorescence colony staining. In: Hampton, R., Ball, E. and De Boer, S. (eds), *Serological Methods for Detection and Identification of Viral and Bacterial Plant Pathogens.* APS Press, St Paul, Minnesota, pp. 299–305.

Van Vuurde, J.W.L. (1990b) Immunostaining of colonies for sensitive detection of viable bacteria in sample extracts and on plant parts. In: Klement, Z. (ed.), *Proceedings of the 7th International Conference on Plant Pathogenic Bacteria, Budapest*, pp. 907–912.

Van Vuurde, J.W.L. and Roozen, N.J.M. (1990) Comparison of immuno-fluorescence colony-staining in media, selective isolation on pectate medium, ELISA and immunofluorescence cell staining for detection of *Erwinia carotovora* subsp. *atroseptica* and *E. chrysanthemi* in cattle manure slurry. *Netherlands Journal of Plant Pathology* 96, 75–89.

Van Vuurde, J.W.L. and Van der Wolf, J.M. (1995) Immunofluorescence colony-staining (IFC). In: Akkermans, A.D.L., Van Elsas, J.D. and De Bruijn, F.J. (eds), *Molecular Microbial Ecology Manual.* Kluwer Academic, Dordrecht, pp. 1–19.

Van Vuurde, J.W.L., Franken, A.A.J.M., Birnbaum, Y. and Jochems, G. (1991) Characterization of immunofluorescence microscopy and of dilution-plating to detect *Pseudomonas syringae* pv. *phaseolicola* in bean seed lots and for risk assessment of field incidence of halo blight. *Netherlands Journal of Plant Pathology* 97, 233–244.

Van Vuurde, J.W.L., De Vries, P.M. and Roozen, N.J.M. (1992) Application of immunofluorescence colony-staining (IFC) for monitoring populations of *Erwinia* spp. on potato tubers, in surface water and in cattle manure slurry. In: Lemattre, M., Freigoun, S., Rudolph, K. and Swings, J.G. (eds), *Proceedings of the 8th International Conference on Plant Pathogenic Bacteria, INRA, Paris*, pp. 741–746.

Van Vuurde, J.W.L., Kastelein, P. and Van der Wolf, J.M. (1995) Immunofluorescence colony-staining (IFC) as a concept for bacterial detection in quality testing of plant materials and ecological research. *EPPO Bulletin* 25, 157–162.

Van Vuurde, J.W.L., De Vries, P.M. and Lopez, M.M. (1994) Colonization of potato stems by *Erwinia* spp., including *in planta* detection in stem sections with immunofluorescence colony-staining (IFC) after *in situ* enrichment in agar medium. In: Lemattre, M., Freigoun, S., Rudolph, K. and Swings, J.G. (eds), *Proceedings of the 8th International Conference on Plant Pathogenic Bacteria, INRA, Paris*, pp. 747–752.

23

Comparison of Serological and Molecular Methods for Detection of *Xanthomonas oryzae* pv. *oryzae* in Rice Seed

A.M. Alvarez[1], F.-U. Rehman[1] and J.E. Leach[2]

[1]Department of Plant Pathology, University of Hawaii, Honolulu, HI 96822; [2]Department of Plant Pathology, Kansas State University, Manhattan, KS 66506, USA

INTRODUCTION

Bacterial blight of rice, caused by *Xanthomonas oryzae* pv. *oryzae*, is a serious disease in rice-growing areas worldwide. The disease is seedborne and hence poses a problem to international seed trade. However, the extent to which the pathogen, once it becomes established in seed, survives, is transmitted and subsequently infects the next generation of emerging rice seedlings is questioned (Mew *et al.*, 1989). The significance of the seedborne phase is difficult to assess because detection of *X. oryzae* pv. *oryzae* in infested rice seed is masked by numerous faster growing rice seed contaminants.

Improved methodology is needed for detection and identification of *X. oryzae* pv. *oryzae* in rice seed. Serological and DNA-based methods have been developed for detection of many seedborne bacterial pathogens including *X. oryzae* pv. *oryzae*. Serological methods include the enzyme-linked immunosorbent assay (ELISA) using monoclonal and/or polyclonal antibodies, miniplate enrichment/ELISA, immunofluorescence and the immunofluorescence colony-staining technique (IFC) (Benedict *et al.*, 1989; Quimio, 1989; Rehman, 1995; Van Vuurde and Van der Wolf, 1995). Current studies focused on IFC to increase the resolution and enhance the sensitivity of previous immunofluorescence detection methods.

 Seed Health Testing (eds J.D. Hutchins and J.C. Reeves)

DEVELOPMENT OF IFC TECHNIQUE FOR *X. ORYZAE* PV. *ORYZAE*

The basic IFC technique (Van Vuurde, 1987; Van Vuurde and Van der Wolf, 1995) was modified to detect and identify *X. oryzae* pv. *oryzae*. Species- and pathovar-specific monoclonal antibodies (mAbs) were generated and characterized using 178 strains assembled from a worldwide collection (Benedict *et al.*, 1989). Most strains reacted with mAb Xco-1 (clone 139-159, IgMκ) but this mAb gave weak immunofluorescence by IFC. In contrast, mAb Xco-2 (clone 138-68, IgG3κ) gave bright immunofluorescence, but it did not react with all strains (Benedict *et al.*, 1989). Additional mAbs were generated to identify nonreactive strains. mAbs of subclass IgG were selected because they diffused through agar and produced brighter colonies than IgM antibodies (Rehman, 1995). A new mAb (Xoo-7, clone 211-G4, IgG3κ) was particularly useful because it reacted with numerous strains from Nepal and some from India that did not react with mAb Xco-2. mAbs Xco-2 and Xoo-7 reacted with 85% and 12%, respectively, of 268 typical strains of *X. oryzae* pv. *oryzae*. The strains that reacted with mAb Xco-2 did not react with mAb Xoo-7 and vice versa. Only eight of the 268 typical strains failed to react with either of these mAbs. A mixture of mAbs Xco-2 and Xoo-7 reacted with 97% of the *X. oryzae* pv. *oryzae* strains that induced typical bacterial blight symptoms in pathogenicity tests.

EVALUATION OF SEMI-SELECTIVE MEDIA

Successful application of IFC depends on rapid growth of target colonies during the enrichment step. Thus, five semi-selective media were evaluated with representative *X. oryzae* pv. *oryzae* cultures from a worldwide collection (Gnanamanickam *et al.*, 1994). Of these, XOS medium, developed by Di *et al.* (1991), was found to be the most suitable. However, typical strains of *X. oryzae* pv. *oryzae* required 6–8 days before colonies could be enumerated under 40× magnification on agar plates. A few strains formed colonies in 5 days, but only atypical fast growing (and often avirulent) strains produced colonies in 3 days. When phosphate was removed from XOS medium, distinct colonies were visible under 40× magnification in 3–4 days and fast-growing strains were visible in 2 days. The efficiency of plating on the modified medium (medium E) was not statistically different from the efficiency on XOS medium. Thus, because of the more rapid colony development, medium E was used for IFC studies.

DETECTION OF *X. ORYZAE* PV. *ORYZAE* IN EXTRACTS OF ARTIFICIALLY INFESTED RICE SEED

Sterile (autoclaved and dried) rice seeds were artificially infested by soaking individual batches of 300 seeds in a 5 ml suspension of *X. oryzae* pv. *oryzae* strain PXO86 (approximately 2×10^8 cfu (colony-forming units) ml^{-1}). Seeds were dried overnight in a laminar flow hood then stored in a dessicator at 4°C. The artificially infested seed was mixed with viable healthy seed at rates of 1, 5, 10 and 100%. Subsamples of 100 infested seed were placed in 200 ml plastic beakers containing 5 ml sterile saline and crushed with a pestle for 3 min. The mixture was incubated for 2 h on a rotary shaker adjusted to 200 rpm. The extract was decanted from the seed and centrifuged at 12,350 *g* for 10 min. The pellet was resuspended in 1 ml saline, ten-fold dilutions were made and 100 µl was added to pour plates. For pour plating, medium E with 1.2% molten agar was cooled to approximately 37°C. A 100 µl aliquot of the appropriate dilutions of the seed extract was placed in the center of a 90 mm sterile Petri plate, and approximately 12 ml of medium E was added per plate. Plates were immediately swirled to spread the suspensions prior to solidification of the agar. Plates were incubated at 28°C and cfu were recorded from day 1 to day 7. The numbers of rice seed contaminants were determined by plating noninfested seed extracts separately on medium E. Dilutions of a pure culture of PXO86 were plated separately on medium E; each sample had four replicates. When pinpoint colonies of *X. oryzae* pv. *oryzae* became visible in medium E under 40× magnification, agar squares (approximately 1 cm^2) were cut from the plate, placed in a microplate (12 wells per plate) and stained with fluorescein isothiocyanate (FITC)-conjugated mAb Xco-2 (1 : 50 dilution) for 12 h. In subsequent assays a mixture of mAb Xco-2 and mAb Xoo-7 (separately conjugated to FITC) were mixed 1 : 1 prior to staining. Colonies showing bright immunofluorescence were lifted with a needle and restreaked on to yeast dextrose calcium carbonate medium (YDC) containing 10 g yeast extract, 20 g dextrose, 20 g $CaCO_3$ and 13 g agar per litre. Colonies were compared with colonies of pure cultures of PXO86, then tested by direct immunofluorescence, and inoculated into leaves of rice cultivar IR-20 for pathogenicity tests using the scissor-clipping method (Kaufman *et al.*, 1973).

Using the IFC technique, pure cultures of *X. oryzae* pv. *oryzae* were detected in approximately 30% of the samples plated at the 10^{-7} dilution (1 cfu per 100 µl sample) and in all samples plated at the 10^{-6} dilution (10 cfu per 100 µl sample). Fluorescent colonies were detected in all replications when healthy seed lots were mixed with infested seed at 1% or greater (Table 23.1). At least one or more colonies showed bright fluorescence even when contaminants outnumbered fluorescent colonies by 37:1. In a repeated experiment, fluorescent colonies were detected in all of five replicate samples with 1% infested rice seeds (as well as all samples with 5% and 100% infested seed). No

Table 23.1. Detection of *Xanthomonas oryzae* pv. *oryzae* (strain PXO86) on medium E (pour plate) from infested rice seed.

Sample no.	Infestation (%)[a]	Cfu ml^{-1}[b]		Reaction in IFC[d]
		Contaminants	*X. oryzae* pv. *oryzae*[c]	
1	0	2.6×10^5	none	–
2	1	3.0×10^5	8.0×10^3	+
3	5	3.3×10^5	2.0×10^4	+
4	10	2.7×10^5	7.4×10^4	+
5	100	none	5.1×10^5	+

[a]Infested seeds per 100 seeds.
[b]Calculated from observed cfu in dilutions of samples 1–5.
[c]Presumptive *Xanthomonas oryzae* pv. *oryzae* based on colony morphology after 3 days' growth.
[d]The immunofluorescence colony staining technique was used to identify *Xanthomonas oryzae* pv. *oryzae* from dried agar pieces (approximately 1 cm^2). A positive reaction is defined as one or more colonies showing bright immunofluorescence.

fluorescent colonies were detected in control samples. Representative fluorescent colonies that were restreaked on to YDC medium appeared identical to colonies of PXO86. They gave bright immunofluorescence when reacted with FITC-conjugated mAb Xco-2 but not mAb Xoo-7. Pathogenicity tests on rice cultivar IR-20 were positive.

Among rice seed contaminants, several species of bacteria that form yellow colonies can be confused with *X. oryzae* pv. *oryzae*. One is a Gram-negative, oxidative bacterium identified by Biolog Microstation System™ as *Pseudomonas paucimobilis*. Another (also Gram-negative, oxidative) resembled *X. campestris* in culture and was identified by Biolog as *X. campestris* pv. *strelitizia*. Although these *Xanthomonas*-like yellow bacteria had other bacteriological characteristics of *X. campestris*, they were nonpathogenic on *Strelitizia reginae* as well as rice. Similar yellow-pigmented, Gram-negative, oxidative bacteria have been reported previously in association with rice seed (Benedict *et al.*, 1989; Di *et al.*, 1991; Gnanamanickam *et al.*, 1994).

ASSESSMENT OF THE IFC TECHNIQUE

The advantages of the IFC technique over previously applied serological methods (ELISA and immunofluorescence) lie in its greater sensitivity (Van Vuurde and Van der Wolf, 1995; Van Vuurde *et al.*, 1995). With IFC lower limits of detection are 1–10 cfu per 100 μl sample (10–100 cfu ml^{-1}) whereas with direct ELISA, strong signals are observed only when bacterial suspensions exceed 5×10^5 cfu (Alvarez and Lou, 1985; Gnanamanickam *et al.*, 1994). Microcolonies observed with IFC are readily detected in a microscope field; thus, larger sample numbers can be handled with relative ease. Moreover, living and dead cells are clearly distinguished because dead cells do not develop into fluorescing microcolonies. Thus, even if stained with the antibody, single cells are not visible in the range of observation (40–200×).

MOLECULAR METHODS FOR DETECTION AND IDENTIFICATION OF *X. ORYZAE* PV. *ORYZAE*

Molecular methods have been useful for identifying *X. oryzae* pv. *oryzae*, analysing natural populations in the Philippines and comparing them with populations worldwide (Leach *et al.*, 1990, 1995). Recently, Vera Cruz *et al.* (1995, 1996) compared Rep-PCR with RFLPs produced by hybridization with IS*1112*, an insertion element isolated from *X. oryzae* pv. *oryzae*. The genetic groups detected by Rep-PCR were consistent with those found by RFLP analysis. Rep-PCR was selected as the method of choice for further population studies because it is less expensive and simpler to perform. An assay called IS-PCR was then developed using outwardly directed primers that amplify DNA between the endogenous *Xoo* insertion elements, IS*1112* and IS*1113*. In an analysis of 300 different strains of *X. oryzae* pv. *oryzae* collected from Nepal and other rice-growing countries around the world, polymorphisms related to geographical origin of the strains were detected and hence were useful in analysis of population structure (Adhikari and Leach, unpublished data; Vera Cruz and Leach, unpublished data). J1, a single outwardly directed primer based on the sequence of IS*1112*, distinguished *X. oryzae* pv. *oryzae* from *X. oryzae* pv. *oryzicola*. Primers N1 and N2, outwardly directed from IS*1113*, generated patterns that distinguished *X. oryzae* pathovars from other pathovars of *X. campestris*. Thus patterns generated by IS-PCR are also useful for diagnostic purposes if two sets of primers are used in combination (Adhikari, Vera Cruz and Leach, unpublished data).

A PCR assay using primers which amplified an internal fragment of the *Xoo* element IS*1112* was developed by Cottyn *et al.* (1994) to detect a single band of *X. oryzae* pv. *oryzae* DNA. This assay showed good potential for detecting *X. oryzae* pv. *oryzae* in pure culture and is presently being tested for rice seed assays (Mew, personal communication). A similar assay, based on the amplification of an internal fragment of the element IS*1113*, was developed by Sakthivel, Nelson and Leach (unpublished data). IS*1113* hybridized to over 1000 strains of *X. oryzae* pv. *oryzae* from a worldwide collection, and hybridized to only a few other xanthomonads; thus, it was thought that primers based on this element might be more specific to *X. oryzae* pv. *oryzae*. After screening several combinations of primers, TXT (5′-GTCAAGCCAACTGTGTA-3′) and

TXT4R (5′-CGTTCGCGCCACAGTTG-3′) were selected (Sakthivel, unpublished data). These primers consistently amplified a 964 bp fragment from *X. oryzae* pv. *oryzae* but not from *X. oryzae* pv. *oryzicola* or pathovars of *X. campestris*, *Pseudomonas* spp. or saprophytic bacteria (Sakthivel and Leach, unpublished data). It should be noted that these primers were 17-mers that had been used for sequencing. We did not try 20-mers which would have increased the specificity and stability of binding.

The PCR assay used in current studies was based on the method of Sakthivel and Leach using primers that amplify a portion of IS*1113*. The reaction mixture (total 25 μl) contained 10 mM Tris (pH 8.3), 1.5 mM $MgCl_2$, 25 μM KCl, 200 μM each of four dNTPs, 0.1 μM forward primer (TXT), 0.1 μM reverse primer (TX4R), 2.5 U *Taq* polymerase and 20 ng template DNA or 2.0 μl test sample. The program was 1 min at 94°C, followed by 30 cycles of 1 min at 94°C, 2 min at 54°C and 2 min at 72°C. PCR products were resolved by electrophoresis in 1.5% agarose gels at 50 V for 2 h using standard procedures.

SENSITIVITY OF PCR USING TXT/TX4R PRIMERS

A ten-fold dilution series of a suspension of strain PXO86 containing 3.5×10^8 cfu ml^{-1} was plated on CPG agar (1.0 g casein, 10.0 g peptone, 10 g glucose, 16 g agar) and cell counts were made after 5 days' growth. One millilitre aliquots of the whole cell preparations were placed into a 90°C water bath for 2 min, and PCR was performed using 2 μl subsamples per PCR reaction (total 25 μl). A clear band formed with dilutions of 10^{-4} whereas only a faint band formed at the 10^{-5} dilution. Based on the dilution series of live *X. oryzae* pv. *oryzae* cells, this dilution contained 35 cfu per 2 μl subsample or 1.6×10^4 cfu ml^{-1}.

Sensitivity was also determined by comparing bands formed by the PCR product from dilutions of DNA (20 ng μl^{-1}) prepared from PXO86. The 10^{-5} dilution of DNA contained approximately 200 fg DNA. Based on the assumption that the genome size of *X. oryzae* pv. *oryzae* is approximately equivalent to the genome size of *Pseudomonas fluorescens* (7.4×10^3 kb) (Bak *et al.*, 1970) the 200 fg DNA represented approximately 27 cfu per 2 μl added to the PCR reaction mix or 1.3×10^4 cfu ml^{-1} in the original sample. Sakthivel reported that the TXT and TX4R primers detected as little as 55 fg (unpublished data).

An attempt was made to monitor the spread of *X. oryzae* pv. *oryzae* strain PXO86 in rice plants and extracts of seeds harvested from infected plants. Results were inconclusive because bacterial DNA was not detected in inoculated rice leaves or in seed extracts. In preliminary experiments using rice plants inoculated with strain PXO99 Sakthivel (unpublished data) detected bacterial DNA in leaves above inoculated leaves collected at 15, 30 and 45 days after inoculation but he did not detect bacterial DNA in seeds harvested from infected plants. Likewise, no bacterial DNA was detected in seedlings germinated from 200 seeds collected from the infected plants. In a preliminary report of these data it was concluded that although the PCR assay can be used to detect the pathogen in infected leaves and stems, more studies are needed to determine whether or not *X. oryzae* pv. *oryzae* is transmitted through the vascular tissue into developing seed (Gnanamanickam *et al.*, 1994).

COMPARISONS BETWEEN THE IFC AND PCR ASSAYS

In the described protocols, the limits of detection of *X. oryzae* pv. *oryzae* are lower with the IFC assay (10 cfu ml^{-1}) than for the PCR assay (1.3×10^4 cfu ml^{-1}). The greater sensitivity of the IFC assay is due primarily to the larger sample size (100 μl) examined per replicate. Given the ease of plating additional replicate samples, the sample size for IFC can be increased even further. Since the sensitivity of the assay is dependent upon the total volume of sample plated, plating ten replicate samples would easily lower the sensitivity to 1 cfu ml^{-1}. With the added advantage that the identity of a fluorescing colony can be confirmed by reisolation and a pathogenicity test, the IFC test is very reliable. Nevertheless, the 2–4 days needed for development of microcolonies prior to staining them with antibody conjugates is a distinct disadvantage of the IFC method. In contrast, the PCR assay can be completed within 1 day. Additional modifications of the DNA-based protocol, including extraction and condensation of target DNA from larger sample volumes, would improve the sensitivity of the PCR assay. On the other hand, use of an immunocapture assay using a mixture of mAbs Xco-2 and Xoo-7 on the solid phase followed by a washing step and PCR would result in a sensitive assay that could be performed in 1 day. The presence of viable cells could be demonstrated later by plating a subsample of the immunocaptured *X. oryzae* pv. *oryzae* cells. Subsequent pathogenicity tests would confirm the identification.

In the case of bacterial blight of rice additional research is clearly needed before the significance of seed transmission is established. Nevertheless, tools are available to examine this question with precision. Extensive field studies and analysis of population structure of *X. oryzae* pv. *oryzae* have formed an essential foundation for detection methods and have revealed polymorphism among strains isolated from representative geographical areas of rice growing regions throughout Asia (Leach *et al.*, 1990, 1992, 1995; Nelson *et al.*, 1994; Vera Cruz *et al.*, 1995, 1996; Ardales *et al.*, 1996). These detailed characterizations of the *X. oryzae* pv. *oryzae* populations and the numerous cultures available for testing are invaluable for validating the DNA-based methods used for detection of this pathogen.

ACKNOWLEDGEMENTS

The authors thank Albert A. Benedict and Carla Mizumoto for collaboration in producing monoclonal antibodies; Natarajan Sakthivel for developing the PCR reaction and sharing unpublished results; Paula Schwenk for able assistance; Casiana Vera Cruz for helpful advice and comments. Research was supported by a grant to A. Alvarez (award no. 92-37311-8295) NRI Competitive Grants Program/USDA.

REFERENCES

Alvarez, A.M. and Lou, K. (1985) Rapid identification of *Xanthomonas campestris* pv. *campestris* by ELISA. *Plant Disease* 69, 1082–1086.

Ardales, E., Leung, H., Vera Cruz, C.M., Mew, T.W., Leach, J.E. and Nelson, R.J. (1996) Hierarchical analysis of spatial variation of the rice bacterial blight pathogen across diverse agroecosystems in the Philippines. *Phytopathology* 86, 241–252.

Bak, A.L., Christiansen, C. and Stenderup, A. (1970) Bacterial genome sizes determined by DNA denaturation studies. *Journal of General Microbiology* 64, 337–380.

Benedict, A.A., Alvarez, A.M., Berestecky, J., Imanaka, W., Mizumoto, C.Y., Pollard, L.W., Mew, T.W. and Gonzalez, C.F. (1989) Pathovar-specific monoclonal antibodies for *Xanthomonas campestris* pv. *oryzae* and for *Xanthomonas campestris* pv. *oryzicola*. *Journal of General Microbiology* 64, 337–380.

Cottyn, B., Bautista, A.T., Nelson, R.J., Leach, J.E., Swings, J. and Mew, T.W. (1994) Polymerase chain reaction amplification of DNA from bacterial pathogens of rice using specific oligonucleotide primers. *International Rice Research Note* 19, 30–32.

Di, M., Ye, H., Schaad, N.W. and Roth, D.A. (1991) Selective recovery of *Xanthomonas* spp. from rice seed. *Phytopathology* 81, 1358–1363.

Gnanamanickam, S.S., Shigaki, T., Medalla, E.S., Mew, T.W. and Alvarez, A.M. (1994) Problems in detection of *Xanthomonas oryzae* pv. *oryzae* in rice seeds and potential for improvement using monoclonal antibodies. *Plant Disease* 78, 173–178.

Kaufman, H.E., Reddy, A.P.K., Hsieh, S.P.Y. and Merca, S.D. (1973) An improved technique for evaluating resistance of rice varieties to *Xanthomonas oryzae*. *Plant Disease Reporter* 57, 537–541.

Leach, J.E., White, F.F., Rhoads, M.L. and Leung, H. (1990) Repetitive DNA sequences differentiate *Xanthomonas campestris* pv. *oryzae* from other pathovars of *X. campestris*. *Molecular Plant–Microbe Interactions* 3, 238–246.

Leach, J.E., Rhoads, M.L., Vera Cruz, C.M., White, F.F., Mew, T.W. and Leung, H. (1992) Assessment of genetic diversity and population structure of *Xanthomonas oryzae* pv. *oryzae* with a repetitive DNA element. *Applied and Environmental Microbiology* 58, 2188–2195.

Leach, J.E., Leung, H., Nelson, R.J. and Mew, T.W. (1995) Population biology of *Xanthomonas oryzae* pv. *oryzae* and approaches to its control. *Current Opinion in Biotechnology* 6, 298–304.

Mew, T.W., Unnamalai, N. and Baraoidan, M.R. (1989) Does rice seed transmit the bacterial blight pathogen? In: Banta, S.J. (ed.), *Bacterial Blight of Rice*. International Rice Research Institute, Manila, The Philippines, pp. 55–63.

Mew, T.W., Alvarez, A.M., Leach, J.E. and Swings, J. (1993) Focus on bacterial blight of rice. *Plant Disease* 77, 5—12.

Nelson, R.N., Baraoidan, M.R., Vera Cruz, C.M., Yap, I.V., Leach, J.E. , Mew, T.W. and Leung, H. (1994) Relationship between phylogeny and pathotype for the bacterial blight pathogen of rice. *Applied and Environmental Microbiology* 60, 3275–3283.

Quimio, A.J. (1989) Serology of *Xanthomonas campestris* pv. *oryzae*. In: Banta, S.J. (ed.) *Bacterial Blight of Rice*. International Rice Research Institute, Manila, The Philippines, pp. 19–30.

Rehman, F. (1995) Serological and pathological evaluation of *Xanthomonas oryzae* pv. *oryzae*, the causal agent of bacterial blight of rice. PhD Dissertation, University of Hawaii, Honolulu, Hawaii, 207 pp.

Van Vuurde, J.W.L. (1987) New approach in detecting phytopathogenic bacteria by combined immunoisolation and immunoidentification assays. *Bulletin OEPP/EPPO* 17, 139–148.

Van Vuurde, J.W.L. and Van der Wolf, J.M. (1995) Immunofluorescence colony-staining. In: Akkermans, A.D.I., Van Elsas, J.D. and De Bruijn, F.J. (eds), *Molecular Microbiology Ecology Manual*. Kluwer Academic, Dordrecht, pp. 1–19.

Van Vuurde, J.W.L., Kastelein, P. and Van der Wolf, J. M. (1995) Immunofluorescence colony-staining (IFC) as a concept for bacterial detection in quality testing of plant materials and ecological research. *Bulletin OEPP/EPPO* 25, 157–162.

Vera Cruz, C.M., Halda-Alija, L., Louws, F.J., Skinner, D.Z., George, M.L., Nelson, R.J., de Bruijn, F.J., Rice, C.W. and Leach, J.E. (1995) Repetitive sequence-based polymerase chain reaction of *Xanthomonas oryzae* pv. *oryzae* and *Pseudomonas* species. *International Rice Research Note* 20, 23–24.

Vera Cruz, C.M., Ardales, E.Y., Skinner, D.Z., Talag, J., Nelson, R.J., Louws, F.J., Leung, H., Mew, T.W. and Leach, J.E. (1996) Measurement of haplotypic variation in *Xanthomonas oryzae* pv. *oryzae* within a single field by Rep-PCR and RFLP analyses. *Phytopathology* 86, 1352–1359.

24

Immunocapture Reverse Transcriptase PCR for the Detection of Lettuce Mosaic Virus

R.A.A. Van der Vlugt[1], M. Berendsen[1] and H. Koenraadt[2]

[1]DLO Research Institute for Plant Protection (IPO-DLO), PO Box 9060, 6700 GW Wageningen; [2]The Netherlands General Inspection Service (NAKG), PO Box 27, 2370 AA Roelofarendsveen, The Netherlands

INTRODUCTION

Lettuce mosaic virus (LMV) is a destructive potyvirus of lettuce (*Lactuca sativa*) and endive (*Cichorium intybus*) with a worldwide distribution. It is transmitted in a non-persistent manner by many aphid species. It is also seedborne (Dinant and Lot, 1992) and can be introduced by LMV-contaminated seed. Therefore the use of virus-free seeds is of major importance to control the spread of the virus.

In the Netherlands ELISA is used to test seed lots for the presence of the virus. In this assay extracts of 2000 seedlings (in 20 subsamples of 100 seedlings each) are tested. In the USA extracts of 30,000 seeds (60 subsamples of 500 seeds) are tested with ELISA in quality control programmes. Currently both tests are performed at the NAKG. Occasionally the interpretation of ELISA test results is difficult because absorption values (at 405 nm) are around the cutoff threshold which is set at 2.5× (for the seedling assay) or 2× (for the seed assay) the average absorption value for the healthy control. Sometimes it is unclear whether low ELISA extinction values are due to the presence of low quantities of LMV or to non-specific binding of the conjugate. Because of this phenomenon there is a possibility of false-negative (virus concentration too low) and false-positive (background too high) test results. A more sensitive method would be useful to determine more accurately the presence of the virus in a subsample. This method could also be used to evaluate the arbitrary thresholds that are used in ELISA.

(eds J.D. Hutchins and J.C. Reeves)

Immunocapture reverse transcriptase polymerase chain reaction (IC RT-PCR) is a method that has been developed for the detection of several viruses with an RNA genome (Nolasco *et al.*, 1993). The method is a combination of techniques that facilitate the detection of viruses present in low concentrations in plant extracts. The first step, immunocapture, is comparable to that of ELISA: virus particles are trapped on the wall of an antiserum-coated tube and inhibitory plant extracts are removed by washing. The viral RNA is released from the particles and used as template for cDNA-synthesis using reverse transcriptase. The resulting cDNA is then amplified in a PCR with virus-specific primers. Finally, analysis of the PCR product is generally performed by electrophoresis on an agarose gel. Here we report on the development of an IC RT-PCR procedure for the sensitive detection of LMV and discuss some factors that can influence the reliable detection of the pathogen in seed lots.

MATERIALS AND METHODS

Virus Isolates

The LMV isolates used in this study were derived from the plant virus collection at IPO-DLO and have been described previously (Bos *et al.*, 1994). LMV-Ls1, LMV-Ls252, LMV-Ls265, LMV-Gr5, LMV-Gal, LMV-F, LMV-W and LMV-Yar were mechanically inoculated and maintained on the susceptible lettuce cultivar Patty as described previously (Bos *et al.*, 1994). LMV was purified according to standard procedures.

Virus Extraction

Samples from lettuce leaves showing clear symptoms (3–4 weeks after inoculation) were ground in phosphate-buffered saline (PBS; 100 mg ml^{-1}) using a stainless steel pestle exactly fitting a 1.5 ml Eppendorf tube. For the seedling assay lettuce seeds were incubated at 10°C for 2 days on moistened filter paper to break dormancy. Then the seeds were incubated at 20°C for 4 days (8 h photoperiod) to induce germination. The seedlings (100 per subsample) were extracted in a roller press using PBST (0.05 M PBS with 0.05% Tween 20, 2% polyvinylpyrrolidone (PVP) and 0.2% bovine serum albumin (BSA)). Seed subsamples (500 seeds) were ground at high speed for 20 s in 5 ml of PBST using an Ultra-turrax.

ELISA

Double Antibody Sandwich (DAS)-ELISA was performed according to standard procedures (Clark and Adams, 1977). LMV antiserum (coating and alkaline phosphatase labelled conjugate) against LMV isolate Ls1 were obtained from IPO-DLO. The sensitivity of the DAS-ELISA was determined using a dilution series of purified LMV in leaf extract of a healthy lettuce plant.

Immunocapture and cDNA Synthesis

Microcentrifuge tubes (0.5 ml) were filled with 50 µl LMV antiserum (1 : 500 in PBS). The tubes were incubated at 37°C for 3 h and rinsed three times with PBS. Next, extracts of leaves, seeds or seedlings (200 µl) were incubated overnight in the tubes at 4°C. Finally the tubes were washed three times with PBS to remove unbound plant extract. After the final wash the tubes were drained of PBS as thoroughly as possible and cDNA synthesis was performed.

Primers

A highly conserved motif in the potyviral coat protein (WCIE/DNG) was used to design the degenerate, potyvirus-specific upstream primer, CPUP. Oligo(dT) primer P9502 has been used previously (unpublished data) to facilitate cDNA synthesis from the poly(A) tail present at the 3′-end of all potyvirus RNA genomes. An alignment of the coat protein C-terminal regions and 3′-non-translated regions (NTRs) of the LMV isolates in this study (unpublished data) was used to design the LMV-specific PCR primer set LMV-UP (upstream) and LMV-DW (downstream). All primers contain an additional *Bam*HI restriction site at their 5′-end to facilitate cloning and analysis of amplicons. Relative positions of primers on the LMV RNA genome are indicated in Fig. 24.1.

cDNA Synthesis

Immunocaptured LMV was incubated at 65°C for 10 min in the presence of downstream primers P9502 or LMV-DW. cDNA synthesis with Moloney murine leukaemia virus (MMLV) reverse transcriptase (Life Technologies) was performed in a 40 µl reaction at 37°C for 1 h according to the manufacturer's instructions.

Polymerase Chain Reaction

Five microlitres from the cDNA reaction mix was amplified according to standard procedures using *Taq* polymerase (HT Biotechnology). CPUP or LMV-UP, and P9502 or LMV-DW were used as upstream and downstream primers,

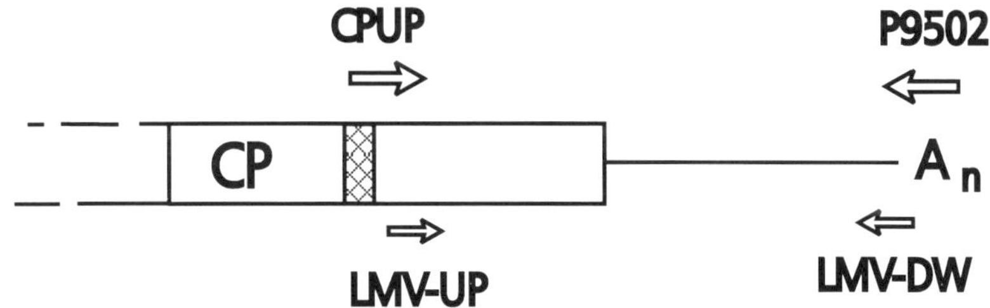

Fig. 24.1. Schematic representation of the 3′-terminal region of the LMV RNA genome. Relative positions of the primers are indicated. Shaded box represents the conserved WCIE/DNG motif in the potyviral coat protein.

respectively. After incubation at 94°C for 5 min the amplification (35 cycles) was performed in a PTC 200 (M.J. Research Inc.) as follows: 94°C for 30 s; 52/57°C for 30 s; and 72°C for 45 s.

Analysis of PCR Products

Amplification reactions were analysed for the expected DNA fragment by agarose gel electrophoresis in 1 × TAE buffer (0.04 M Tris-acetate; 0.001 M EDTA). The DNA in the gel was stained with ethidium bromide and visualized using a UV-light source.

RESULTS

Optimization of cDNA Synthesis

cDNA was synthesized under standard conditions using either oligo(dT) primer P9502 or LMV-specific primer LMV-DW. Subsequent amplification of the cDNAs with the potyvirus-specific primer set P9502/CPUP clearly revealed that cDNA synthesis was more efficient using the oligo(dT)-based primer P9502. Extension of the cDNA reaction time to 2.5 h or raising the temperature of the reaction to 42°C had relatively little effect on the cDNA yield.

Optimization of PCR Conditions

PCR with different primer combinations was performed to determine the optimal primer combination for amplification of oligo(dT)-generated cDNA. All primer combinations (P9502/CPUP, P9502/LMV-UP, LMV-DW/CPUP and LMV-DW/LMV-UP) were tested at either 52°C or 57°C. At both temperatures all primer combinations yielded an amplification product of the expected size (results not shown). However, the yield was clearly dependent on the optimal calculated annealing temperature of the primers involved: 52°C for potyvirus-specific primers P9502 and CPUP, and 57°C for for LMV-specific primers LMV-UP and LMV-DW. The homologous primer combinations yielded more product than the heterologous combinations at their optimal annealing temperatures. Little difference in amplicon yield was observed between primer combinations CPUP/P9502 and LMV-UP/LMV-DW. For reasons of specificity of the detection it was decided to use the LMV-specific primer combination LMV-UP/LMV-DW in further experiments.

Sensitivity of the IC RT-PCR

To compare the sensitivity of the IC RT-PCR with the standard ELISA a dilution series of purified LMV isolate Ls1 in extract of healthy lettuce plants was used in both a standard ELISA test and an IC RT-PCR as described above. The results, summarized in Table 24.1 show that amounts of LMV below 50 ng can no longer be reliably detected by the ELISA format used. However, IC RT-PCR

Table 24.1. Detection of different amounts of purified LMV in healthy leaf extract using ELISA and IC RT-PCR.

Virus concentration (ng)	ELISA value (A_{405} after 1.5 h)	IC RT-PCR
400	1.171	NT
200	0.482	NT
100	0.344	NT
50	0.272	+
5	0.087*	+
0.5	0.072*	+
0.05	0.072*	+
0	0.072	–

+, Presence of LMV-specific DNA product after agarose electrophoresis; –, no PCR product; NT, not tested.
*Below detection threshold.

allows the detection of as little as 50 pg LMV in leaf extract, demonstrating that the IC RT-PCR can be at least a thousand times more sensitive than ELISA.

Specificity of IC RT-PCR

Previous experiments all employed the LMV type isolate Ls1. However, a considerable number of other LMV isolates occur worldwide and it is of utmost importance that all these isolates are also reliably detected. Therefore the specificity of the IC RT-PCR method was tested on different isolates of LMV from the IPO-DLO virus collection. Virus was immunocaptured from lettuce leaves infected with isolate Ls1, Ls252, Gr5, 265, F, Gal, W and Yar (Bos *et al.*, 1994) as described above and subjected to RT-PCR. Figure 24.2 shows that a similar-sized DNA-fragment can be amplified from all LMV isolates. This shows that the antiserum and the primers used in this procedure recognize all LMV isolates involved in this study and that this method can be used to detect different isolates of LMV.

Detection of LMV in Seed Lots

At the NAKG, lettuce seed lots are routinely tested for the presence of LMV by ELISA. Samples of both seed and seedling extracts that were tested in ELISA were also subjected to the IC RT-PCR. For practical reasons, however, samples had to be stored at –20°C before immunocapture. In contrast to the immunocapture on fresh leaf material this sometimes resulted in much lower detection levels of LMV; moreover, the results did not always fully correlate with those of the ELISA. Samples that were clearly positive in ELISA did not always show an amplicon, or showed only a very faint one, after IC RT-PCR (data not shown). Extracts from virus-infected lettuce leaf material that were stored for more than 48 h at –20°C or 4°C always reacted positive both in ELISA and in the IC

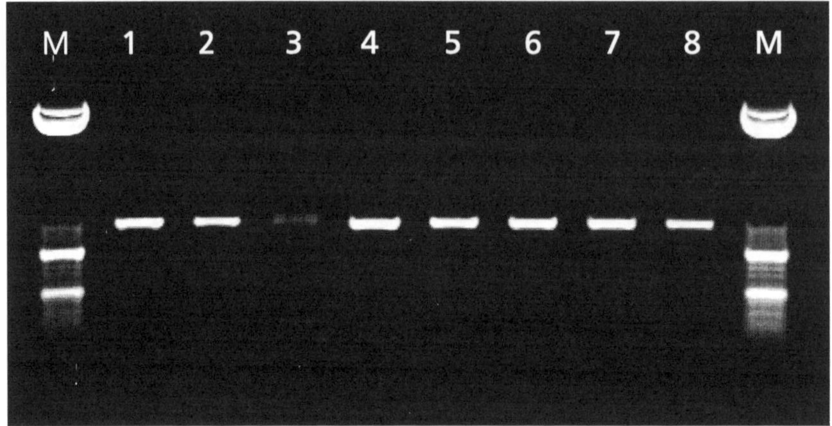

Fig. 24.2. Detection of several LMV isolates in leaf extract of virus-infected lettuce using IC RT-PCR. Lane 1, LMV-Ls1; lane 2, LMV-Ls252; lane 3, LMV-Gr5; lane 4, LMV-Ls265; lane 5, LMV-F; lane 6, LMV-Gal; lane 7, LMV-W; lane 8, LMV-Yar. M, DNA marker (Boehringer Mannheim DNA molecular weight marker XIII; 50 bp ladder).

RT-PCR. This indicates that the seed and seedling extract probably contain components that influence the reliable detection of LMV by IC RT-PCR.

CONCLUSIONS

It was shown that IC RT-PCR can be used to detect small amounts of LMV in leaf extracts. It was approximately a thousand times more sensitive than the current ELISA protocol. The use of LMV-specific primers in the PCR, in combination with a specific antiserum in the immunocapture, will ensure a high specificity of the detection method. In addition the primer set developed still allows the detection of a number of different LMV isolates from different geographic origins.

There are, however, a number of potential problems associated with the detection of LMV in seed samples by the IC RT-PCR method. Sample preparation needs much more attention than with normal ELISA. Due to its high sensitivity, in comparison with ELISA, special precautions need to be taken to prevent cross-contamination of subsequent (healthy) samples by a preceding infected sample. This would result in false-positive samples.

Another problem is that of false-negative samples. Components in the seed and seedling samples appear to inhibit the reproducibility of a sensitive detection of LMV. At the NAKG the IC RT-PCR method was initially intended to retest samples with an ELISA value between the healthy control and the arbitrary threshold of 2.5 times the healthy control for an infected sample. The IC

RT-PCR method holds the potential to be used to determine if these samples are truly infected with (low levels of) LMV. However, additional research will be needed to improve the reproducibility of the method.

REFERENCES

Bos, L., Huijberts, N. and Cuperus, C. (1994) Further observations on variation of lettuce mosaic virus in relation to lettuce (*Lactuca sativa*), and a discussion of resistance terminology. *European Journal of Plant Pathology* 110, 293–314.

Clark, M.F. and Adams, A.N. (1977) Characteristics of the microplate method of enzyme-linked immunosorbent assay for the detection of plant viruses. *Journal of General Virology* 34, 475–483.

Dinant, S. and Lot, H. (1992) Lettuce mosaic virus. *Plant Pathology* 41, 528–542.

Nolasco, G., de Blas, C., Torres, V. and Ponz, F. (1993) A method combining immunocapture and PCR amplification in a microtiter plate for the detection of plant viruses and subviral pathogens. *Journal of Virological Methods* 45, 201–218.

25 Comparison of Immunocapture PCR and ELISA in Quality Control of Pea Seed for Pea Seedborne Mosaic Potyvirus

T.T.H. Phan[1], R.K. Khetarpal[2], T.A.H. Le[1] and Y. Maury[3]

[1]*Institute of Agricultural Genetics, Hanoi, Vietnam;* [2]*National Bureau of Plant Genetic Resources, New Delhi, India;* [3]*Institut National de Recherches Agronomiques, Versailles, France*

INTRODUCTION

Pea seedborne mosaic potyvirus (PSbMV) is a flexuous, rod-shaped virus composed of a 33 kDa capsid protein and a positive-sense, single-stranded RNA molecule 9924 nucleotides in length (Johansen *et al.*, 1991). It is transmitted in pea seeds at high frequencies and its economic importance is linked to this biological property (Maury and Khetarpal, 1992). Seed transmission occurs when the mother plant is infected early in the growing season and during flowering (Wang and Maule, 1992). In each agricultural area, natural seed transmission thus varies with the climatic conditions which govern the intensity of aphid vector flights (Zimmer and Lamb, 1993). As the natural host range of PSbMV is very restricted, the incidence of disease depends on the infection of seed lots. Quality control of seed therefore assumes great importance for controlling this disease.

A simple and practical strategy of quality control of pea seeds for PSbMV has been proposed recently. It is based on the seed transmission property of the virus through the seed lot, implying that the virus in seed coats is not taken into consideration. This quality control involves group testing of seeds in ELISA by taking into account the following points:

1. The production of an antiserum that differentiates between the seed transmitted virus of infected embryos and the non-transmitted virus of infected testas. This specificity resulted from the fact that the virus capsid protein is intact in embryo and truncated in mature testa. The polyclonal antiserum (named antiserum K) obtained from early bleedings detects the cleavage products of the capsid protein and hence does not detect the virus in the testa (Masmoudi *et al.*, 1994a). This permits dispensing with the tedious step of decortication of pea seeds before conducting the test.

2. The probability (Ps1) of detecting one infected embryo in a group of seeds according to the number of seeds in each group. This probability can be experimentally determined for each antiserum (Masmoudi *et al.*, 1994b).

3. The statistical determination of the number of groups to be tested in order to know if the seed lot is at a non-tolerable level of infection or not (Masmoudi *et al.*, 1994b; see Chapter 32).

A way to improve this methodology further would be by choosing a more sensitive technique for detecting PSbMV. Its adoption in quality control would increase the values of Ps1 and further lower the number of groups to be tested.

Nowadays, PCR is recognized as being the most sensitive technique for detecting plant viruses. Its use in a seed test would begin with the extraction of RNA from a large number of seeds, which would be inconvenient in routine use. A variant of PCR known as immunocapture PCR (IC-PCR) has the inherent simplicity of amplifying sequences of viral RNA without prior viral RNA extraction, which would presumably facilitate its use in routine testing. Moreover, because antibodies are involved in the first step of IC-PCR, it may be assumed that, by adopting antiserum K, this method could also selectively detect PSbMV in infected embryos. Therefore, the objectives of the present investigations were to adapt IC-PCR for detecting PSbMV, to evaluate the specific detection property of antiserum K when used in IC-PCR and to check the potential of IC-PCR as compared to ELISA for carrying out a quality control of seeds for viruses in routine.

MATERIALS AND METHODS

IC-PCR Protocol

The IC-PCR protocol developed by Wetzel *et al.* (1992) was adapted for detecting PSbMV both in seeds and leaves of pea. For immunocapture of the virus, Eppendorf tubes were coated separately with two different IgGs, to PSbMV, namely IgG ‘H’ and IgG ‘K’. ‘H’ is a hyperimmune antiserum (Hamilton and Nichols, 1978) to PSbMV kindly given by Dr R.I. Hamilton, Pacific Agriculture Research Center, Agriculture and Agri-food Canada, Vancouver, Canada. ‘K’ is an antiserum to the PSbMV capsid protein (Masmoudi *et al.*, 1994a). Each IgG was used at different dilutions (1–4 $\mu g\ ml^{-1}$) in 100 μl of the coating buffer (carbonate buffer, pH 9.6) and incubated at 36°C for 2–3 h.

The tubes were then rinsed once with PBST (phosphate-buffered saline containing 0.05% Tween) before adding the extracts of seed (or leaf) samples known to be infected by different isolates of PSbMV. The tubes with sample extracts were incubated overnight at 4°C and then washed twice with PBST.

Decapsidation of the immunocaptured viral particles, reverse transcription and PCR were performed in the same tube. Decapsidation was achieved by adding Triton X-100 to the reaction mixture. Variable combinations and concentrations of reverse transcriptase and *Taq* polymerase enzymes manufactured by two and three different companies, respectively, were evaluated. The optimum concentrations of enzymes were then used for all further experiments. Two different sets of primers (D1,D2 and F1,F2) were used, the sequences of which are given below:

D1 (34-mer): 5′-GCT-CTA-GAC-TCG-AGG-GGA-ART-CRA-AAG-CTA-AAA-C-3′
D2 (28-mer): 5′-GTC-CTA-GAG-CTT-GCG-CAA-TWG-GAT-TGT-A-3′

F1 (19-mer): 5′-GAT-TTC-TTC-GTT-GTT-TGT-T-3′
F2 (19-mer): 5′-CTT-GAG-TGC-TGG-CGT-GGT-T-3′

The primers D1 and D2 were suggested by E. Johansen (Denmark). These were degenerated primers corresponding to the VPg-NIa region of the PSbMV genome and gave a 654 bp amplification product. The F1 and F2 primers were obtained by the computer program Oligo-4. They also corresponded to the VPg-NIa region of the genome and gave a 494 bp amplification product.

The final reaction mixture consisted of a solution comprising a mixture of 6 μl of Triton X-100 (1.7%), 5 μl of 10× PCR buffer, 0.5 μl of dNTPs (10 μM), 0.5 μl of primer 1 (100 μM), 0.5 μl of primer 2 (100 μM), 5 units of AMV reverse transcriptase (Pharmacia) and 0.5 units of *Taq* DNA polymerase (Appligene). To this mixture sterilized water was added to make the final volume 50 μl.

The tubes containing the reaction mixture were either directly subjected to PCR or were incubated for 20–30 min in a refrigerator (to allow time for the Triton X-100 to decapsidate the virus) prior to PCR. The PCR cycles (performed in a PTC 200 thermocycler) were: 15 min at 42°C (reverse transcription), 5 min at 92°C (denaturation) with 40 cycles of 20 s at 92°C (denaturation), 20 s at 55°C (annealing of primers), 40 s at 72°C (polymerization) followed by 10 min at 72°C (termination). The amplified DNA was separated on a 1.5% agarose gel along with the 1 kb marker ladder or 123 bp BRL (GIBCO BRL) in TBE buffer at 100 Volts for about 30 min. The DNA bands were visualized by ethidium bromide fluorescence.

Comparing Sensitivity of Detection of PSbMV by IC-PCR and ELISA in Seed Extracts

In order to compare the sensitivity of IC-PCR with DAS-ELISA 30 pea seeds from infected (15% infection rate) and healthy seed lots of pea cv. Belinda were soaked separately overnight in water. After manual dissection, testas and embryos were individually ground in 15 volumes of PBST using a Waring blender, and each extract was centrifuged (10 min, 10,000 *g*). After centrifugation of each seed extract, supernatant of the infected samples was diluted in series in a similar extract from healthy seeds in order to obtain 1 : 30, 1 : 50, 1 : 100, 1 : 1000, 1 : 5000 and 1 : 10,000 dilutions. These diluted extracts were used for detection of the virus both in IC-PCR and in DAS-ELISA.

DAS-ELISA was used as described by Clark and Adams (1977). The 'H' and 'K' immunoglobulins were used at 1 μg ml^{-1} on separate CML M29L microplates and the alkaline phosphatase conjugates at 1/1000 dilution. Ten healthy homogenates were also deposited in duplicate wells of each microplate for the determination of the positive–negative threshold which was defined by a 0.005 probability for a healthy sample to exceed it (Maury *et al.*, 1987).

RESULTS

Adaptation of IC-PCR for Detecting PSbMV

It was observed that for immunocapture of the virus, IgG at a concentration of 3–4 μg ml^{-1} was found to be the most effective. Of the different combinations of enzymes studied, AMV reverse transcriptase (Pharmacia) and *Taq* polymerase (Appligene) used at the rate of 5 units and 0.5 units, respectively, was the most optimum combination. Incubation of the reaction mixture prior to PCR was found to be the most reliable for decapsidation of the virus.

On adapting the IC-PCR protocol, PSbMV was detected in both infected leaf and seed extracts of pea with the two different sets of primers (Fig. 25.1). The primers used in IC-PCR revealed strain specificity in detection, for example, PSbMV isolate Pi7 could not be detected by primers F1 and F2 whereas it was easily detected by primers D1 and D2. However, even on adopting the standardized IC-PCR protocol, the technique was not 100% reproducible. This remains unexplained at present.

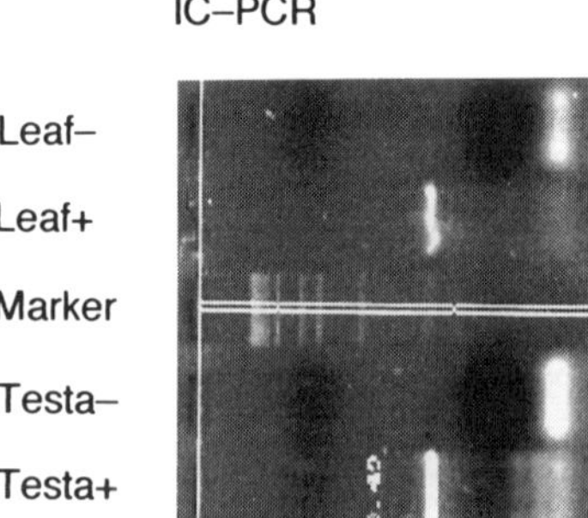

Fig. 25.1. Detection of PSbMV by IC-PCR in leaves and testas of pea. PSbMV was immunocaptured by IgG 'H'. The primers used were F1 and F2. A 1 kb marker ladder was used.

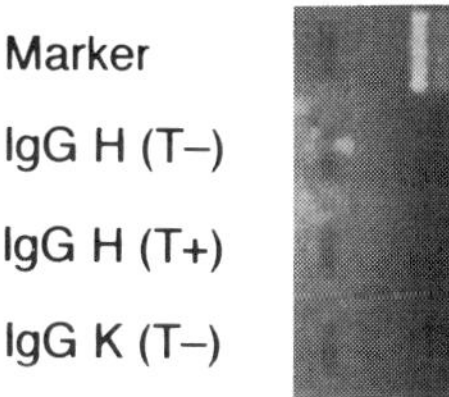

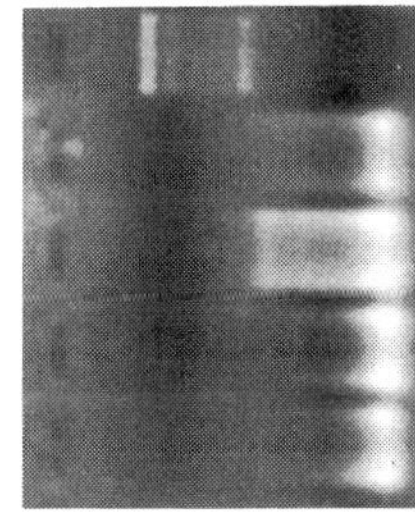

Fig. 25.2. IC-PCR detection of PSbMV in seed testas by using IgG 'H'. Note that IgG 'K' failed to immunocapture the virus.

Evaluation of the Specific Detection Properties of Antiserum 'K' when used in IC-PCR

When IgG 'K' was used for immunocapture in IC-PCR, PSbMV was not detected in the testas whereas it was detected by IgG 'H' (Fig. 25.2). This result indicates that no traces of viral RNA remain in any residual intact capsid in the seed testas.

Comparing Sensitivity of IC-PCR and ELISA for Detecting PSbMV in Seed

When comparing the sensitivity of detection of PSbMV in single infected embryos ground in PBST, preliminary observations revealed that ELISA could detect virus up to a dilution of 1 : 1000 in seed extracts but using IC-PCR the virus could only be detected in a dilution of up to 1 : 30 (Fig. 25.3).

DISCUSSION

Attempts were made to adapt the protocol of IC-PCR given by Wetzel *et al.* (1992) for detecting PSbMV in seeds and leaves of pea. However, there was an inconsistency in reproducibility of results using this technique. This lack of reliability has also been reported for the detection of other plant viruses (Barbara *et al.*, 1995; Astier-Manifacier and Kerlan, personal communication).

IC-PCR was less sensitive than ELISA in detecting the virus in seed; no conclusive reasons for this could be attributed. It is suspected that the presence of certain inhibitors in the seed extracts may be responsible for lowering the sensitivity of IC-PCR. Moreover, due to the low sensitivity observed for IC-PCR, it is logical that only virus present in the embryos can be detected as is the case when using ELISA. The strain specificity of detection by certain primers necessitates the selection of universal primers and the development of multiplex PCR. Such specificity has hardly been observed in DAS-ELISA. This indicates that by using a single set of primers in IC-PCR there are more chances of some of the isolates or a hitherto unknown strain of the virus escaping detection, compared with the classical ELISA test, even on using antiserum K.

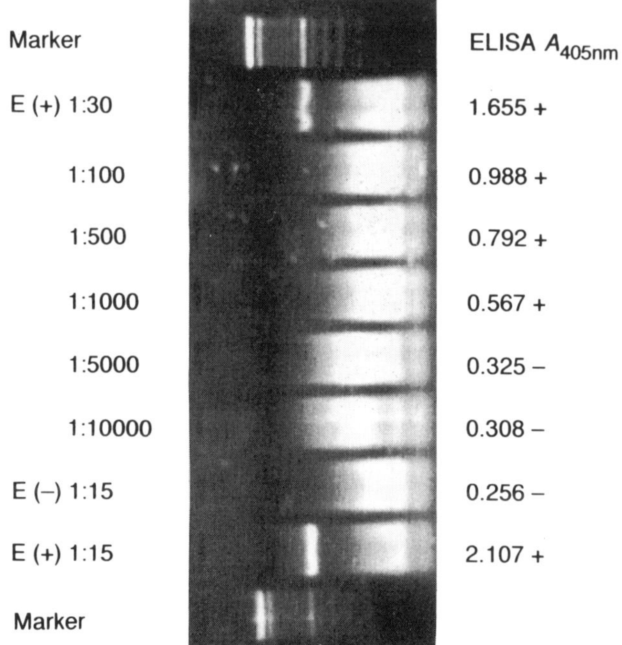

Fig. 25.3. Dilution limit of detection of PSbMV in groups of 40 embryos using IC-PCR and ELISA. PSbMV was immunocaptured by IgG 'K' in both the techniques. The primers used were D1 and D2. The size standard was marker 123 (BRL).

Therefore, it was concluded that IC-PCR would need to be adapted further. Under the present conditions, an improvement in sensitivity can be achieved by performing PCR after extraction of RNA from each sample of seeds. This, however, is a laborious and time-consuming step for use in routine tests; moreover, nucleic acid extraction of a sample cannot be exploited for simultaneous detection of seed-transmitted bacterial and fungal pathogens, because progress in the latter field is via BIO-PCR, which detects the pathogen after growth enrichment on a culture medium (see Chapter 21). Infection of the embryo (except for tobamoviruses) for seed transmission of viruses and the inability for enrichment on culture medium thus make the detection of viruses a different problem. Even if BIO-PCR is applied in routine use for detecting bacteria and fungi, PCR appears less attractive for detection of viruses for which the ELISA methodology has been developed for routine application. Finally, since the technique of PCR has been patented by a private company, its cost-effectiveness for large-scale routine use by seed-testing stations needs to be evaluated.

REFERENCES

Barbara, D.J., Morton, A., Spence, N.J. and Miller, A. (1995) Rapid differentiation of closely related isolates of two plant viruses by polymerase chain reaction and restriction fragment length polymorphism analysis. *Journal of Virological Methods* 55, 121–131.

Clark, M.F. and Adams, A.N. (1977) Characteristics of the microplate method of enzyme-linked immunosorbent assay (ELISA) for detection of plant viruses. *Journal of General Virology* 34, 475–483.

Hamilton, R.I. and Nichols, C. (1978) Serological methods for detection of pea seed borne mosaic virus in leaves and seeds of *Pisum sativum*. *Phytopathology* 68, 539–543.

Johansen, E., Rasmussen, O.F., Heide, M. and Borkhardt, B. (1991) The complete nucleotide sequence of pea seed-borne mosaic virus RNA. *Journal of General Virology* 72, 2625–2632.

Masmoudi, K., Suhas, M., Khetarpal, R.K. and Maury, Y. (1994a) Specific serological detection of the transmissible virus in pea seed infected by pea seed-borne mosaic virus. *Phytopathology* 84, 756–760.

Masmoudi, K., Duby, C., Suhas, M., Guo, J.Q., Guyot, L., Olivier, V., Taylor, J. and Maury, Y. (1994b) Quality control of pea seed for pea seed borne mosaic virus. *Seed Science and Technology* 22, 407–414.

Maury, Y. and Khetarpal, R.K. (1992) Pea seed borne mosaic virus. In: Chaube, H.S., Kumar, J., Mukhopadhyay, A.N. and Singh, U.S. (eds) *Plant Diseases of International Importance*, Volume 2. Prentice Hall, pp. 74–92.

Maury, Y., Bossennec, J.M., Boudazin, G., Hampton, R.O., Pietersen, G. and Maguire, J.D. (1987) Factors influencing ELISA evaluation of transmission of pea seed borne mosaic virus in infected pea seed: seed group size and seed decortication. *Agronomie* 7, 225–230.

Wang, D. and Maule, A.J. (1992) Early embryo invasion as a determinant in pea of the seed transmission of pea seed borne mosaic virus. *Journal of General Virology* 73, 1615–1620.

Wetzel, T., Candresse, T., Macquaire, G., Ravelonandro, M. and Dunez, J. (1992) A highly sensitive polymerase chain reaction method for plum pox potyvirus detection. *Journal of Virological Methods* 33, 355–365.

Zimmer, R.C. and Lamb, R.J. (1993) Amplification and spread of pea seed-borne mosaic virus in field-grown peas. *Canadian Journal of Plant Pathology* 15, 17–22.

26

ISTA/PDC Bacteriology Working Group: Comparative Tests and Working Sheets

N.W. Schaad

ARS-USDA, Foreign Disease–Weed Science, Research Unit, Fort Detrick, MD 21702-5023, USA

The ISTA-PDC Bacteriology Working Group, chaired by N.W. Schaad and assisted by S.J. Roberts and W. Cooper, is currently organized into 15 sub-Working Groups. New subgroups are formed based on interest in organizing a comparative test for a specific pathogen (Table 26.1). The following is a brief summary of the protocol established by the Bacteriology Working Group for comparative testing:

1. The group meets during Workshop every 3 years to propose new tests, or members request a test by corresponding with the Working Group leader. Only assay methods published in peer-reviewed journals are acceptable for comparative testing.
2. The subgroup leader identifies up to five members wishing to participate and appoints an organizer. The organizer prepares an outline of the protocol and data sheet and sends these with a copy of the reviewed publication to the Working Group leader for approval.
3. The organizer obtains a sufficient quantity of naturally contaminated seed for the test and does a pre-test assay to determine the ratio of contaminated to pathogen-free seed necessary to create artificially the three levels of contamination required, i.e. low (some replicates must be negative), moderate (most or all of the replicates positive) and negative (no pathogen detected).
4. The identified seed is divided into four replicates (generally 3000 seeds each) for each level, for each participant.
5. The standard record sheets must be supplied by the organizer, including spaces for positive and negative controls.

Table 26.1. Sub-Working Groups of ISTA-PDC Bacteriology Working Group.

Pathogen	Leader
Acidovorax avenae subsp. *citrulli*	Randhawa
Clavibacter michiganensis subsp. *michiganensis*	Bolkan
Erwinia stewartii	Iles
Pseudomonas syringae	
pv. *glycinea*	Caffier
pv. *lachrymans*	Saygili
pv. *phaseolicola*	Schaad
pv. *pisi*	Roberts
pv. *syringae*	Serfontein
pv. *tomato*	Bolkan and Alvarez
Xanthomonas campestris	
pv. *campestris*	Manceau
pv. *carotae*	Caffier
pv. *holicola*	Claflin
pv. *malvacearum*	Rudolph
pv. *manihotis*	Fessehaie
pv. *vesicatoria*	Sheppard and Koenraadt
pv. *phaseoli*	Bolkan and Alvarez
Xanthomonas oryzae pv. *oryzae* and *oryzicola*	Mortensen and Alvarez

6. The organizer has responsibility for processing the data statistically and must send a copy of the report to the Working Group leader prior to the next Working Group meeting.

Several comparative tests were conducted in 1991–1996; the following were presented at our Workshop in Angers, France in 1992: (i) agar plating test for *Clavibacter michiganensis* subsp. *michiganensis* (H. Bolkan, C.M. Waters and M. Fatmi, USA); (ii) agar plating phaseolotoxin test for *Pseudomonas syringae* pv. *phaseolicola* (H. Jansing and K. Rudolph, Germany); (iii) improved agar plating test for *Xanthomonas campestris* pv. *campestris* (N.W. Schaad, USA, and A. Franken, The Netherlands); (iv) agar plating test for *X. campestris* pv. *translucens* (R.L. Forster and N.W. Schaad, USA).

Several tests organized during the 1992 workshop were presented during our 1996 Bacteriology Workshop in Cambridge. These include: (i) an ELISA test kit for *Erwinia stewartii* (D. McGee and A. Iles, USA); (ii) agar plating test for *X. campestris* pv. *phaseoli* (J. Sheppard, Canada); and (iii) BIO-PCR test for *P. syringae* pv. *phaseolicola* (N.W. Schaad and E. Hatziloukas, USA). A fourth test on BIO-PCR for *Clavibacter michiganensis* subsp. *michiganensis* was organized by H. Bolkan and C. Waters, USA, but was delayed due to the lack of sufficient naturally contaminated seeds.

The following tests were discussed during the 1992 meeting: (i) serological dip stick test kit for *P. syringae* pv. *pisi* (J. Taylor, UK); (ii) serological test for

Table 26.2. Working Sheets published by the ISTA-PDC Bacteriology Working Group in 1996.

Pathogen	Host	Author	No.	Edition
Clavibacter michiganensis subsp. *michiganensis*	Solanaceae	Bolkan, Waters and Fatmi	67	1st
Xanthomonas campestris pv. *translucens*	Poaceae	Forster and Schaad	68	1st
Pseudomonas syringae pv. *phaseolicola*	*Phaseolus vulgaris*	Jansing and Rudolph	66	2nd
Xanthomonas campestris pv. *campestris*	Cruciferae	Schaad and Franken	50	2nd

Xanthomonas holicola (L. Claflin, USA); and (iii) phage test for rapid identification of *Xanthomonas translucans* (R.L. Forster, USA). These tests were organized after the 1996 workshop due to delays in getting the methods published.

During 1996 new Working Sheets were published on *C. michiganensis* subsp. *michiganensis* and *X. campestris* pv. *translucens* and second edition sheets were published on *P. syringae* pv. *phaseolicola* and *X. campestris* pv. *campestris* (Table 26.2).

Comparative Test for the Detection of *Xanthomonas campestris* pv. *campestris* in Crucifer Seeds

27

H. Koenraadt

The Netherlands General Inspection Service for Flower and Vegetable Seeds (NAKG), PO Box 27, 2370 AA Roelofarendsveen, The Netherlands

INTRODUCTION

Xanthomonas campestris pv. *campestris* is an important seedborne pathogen and the causal organism of black rot of crucifers. The use of healthy seed is critical in the control of this bacterial disease. Several methods have been developed to detect *X. campestris* pv. *campestris* on seed. A commonly used method is the seed washing liquid plating assay. In this assay bacteria are first extracted from seed, then dilution plating of the bacterial extract on semi-selective media is used to detect *X. campestris* pv. *campestris* colonies. Consecutively, putative *X. campestris* pv. *campestris* colonies are tested in a pathogenicity assay. Finally the emergence of typical black rot symptoms in the pathogenicity assay indicates that the seed lot is contaminated with *X. campestris* pv. *campestris*.

A number of extraction procedures and semi-selective media are used to detect *X. campestris* pv. *campestris* on seed. A variety of pathogenicity assays are used to identify pathogenic *X. campestris* pv. *campestris* colonies (Schaad and Kendrick, 1975; Randhawa and Schaad, 1984; Yuen *et al.*, 1987; Schaad, 1989; Chang *et al.*, 1991; Schaad and Franken, 1996). It is not yet clear whether certain semi-selective media and pathogenicity assays will give more reliable results than others. In the framework of the International Seed Health Initiative (ISHI) (see Chapter 12) a comparative test to detect *X. campestris* pv. *campestris* on seed of cabbage was organized. The primary goal of the test was to evaluate four semi-selective media and two pathogenicity assays that are used by laboratories in the USA, France and The Netherlands.

MATERIALS AND METHODS

Seed Lots

A number of naturally *X. campestris* pv. *campestris*-contaminated and healthy seed lots were provided by seed companies. In a preliminary test several seed lots were tested for *X. campestris* pv. *campestris*. Eight seed lots with variable numbers of *X. campestris* pv. *campestris* and saprophytes were selected (each sample contained 65,000 seeds) and distributed among 13 participants in the USA, France, and The Netherlands.

Extraction of Bacteria

Three subsamples of 10,000 seeds were transferred to sterile Erlenmeyer flasks. Then 100 ml of sterile 0.85% NaCl was added to each flask. The flasks were incubated on a shaker (250 rpm) at room temperature for 2.5 h. One millilitre of extract was removed from each flask after 5 min and 2.5 h and used for dilution plating.

Dilution Plating

Semi-selective (CS20ABN, mFS, NSCA and NSCAA) media were prepared according to standard procedures (Schaad and Kendrick, 1975; Randhawa and Schaad, 1984; Yuen *et al.*, 1987; Chang *et al.*, 1991). The extract was used to prepare two ten-fold dilutions (10× and 100×) from the 5 min and 2.5 h samples using sterile saline. Then 100 μl of the undiluted, 10×, and 100× dilutions were plated in duplicate on each of the semi-selective media. The media were incubated at 26–28°C for 4 days (CS20ABN, NSCA and NSCAA) to 7 days (mFS). The Petri dishes were evaluated for the presence of putative *X. campestris* pv. *campestris* and saprophytic colonies. The numbers were counted (< 100 colonies per Petri dish) or estimated (> 100 colonies per Petri dish). Putative *X. campestris* pv. *campestris* colonies (maximum 36 colonies/sample) were transferred from the Petri dishes with semi-selective media to yeast extract–dextrose–$CaCO_3$ (YDC) (Wilson *et al.*, 1967). Colonies on YDC with a typical *X. campestris* pv. *campestris* morphology were then tested in two pathogenicity assays.

Pathogenicity Tests

A selection of the YDC-positive colonies (maximum 24 per sample) were tested in the 'excised cotyledon' and the 'cabbage plant' assay. In the excised cotyledon assay one of the cotyledons was removed from 8–10 day old cabbage seedlings with a sharp pair of scissors. Then the fresh wound was immediately inoculated with a cotton swab that had been dipped in a putative *X. campestris* pv. *campestris* colony. In the cabbage plant assay the first and second leaves of 14–21 day old plant were inoculated at four sites. Inoculation was performed by puncturing the secondary veins with a toothpick that had been dipped in a YDC-positive colony. In each assay five plants were inoculated per YDC-positive colony. Both assays were incubated in the greenhouse for 14 days at 25°C. Symptoms were evaluated after 1 and 2 weeks. Symptoms were compared with four reference strains, including non-pathogenic *Xanthomonas* and *X. campestris* pv. *amoraciae* that were included in the pathogenicity assays.

Data Collection

Each participant collected the dilution plating and pathogenicity assay data on pre-prepared record sheets. Statistical analysis is still ongoing.

RESULTS AND DISCUSSION

Eight samples were included in this comparative test. Large differences in the number of saprophytic and putative *X. campestris* pv. *campestris* colonies per sample were observed by all laboratories. Most laboratories detected pathogenic *X. campestris* pv. *campestris* colonies in seven out of eight seed lots. In all laboratories the number of saprophytic and putative *X. campestris* pv. *campestris* colonies in the dilution plating was significantly higher after 2.5 h of extraction. The increase in the number of *X. campestris* pv. *campestris* colonies after 2.5 h of shaking suggests that the longer extraction period will facilitate the detection of *X. campestris* pv. *campestris*. However, due to overgrowth of *X. campestris* pv. *campestris* by saprophytes this was not the case in all samples. In seed lots with an abundant saprophytic microbiota several laboratories were not consistently able to isolate putative *X. campestris* pv. *campestris* colonies. The selectivity of the four media determined to a large extent whether putative *X. campestris* pv. *campestris* colonies could be isolated after 5 min and 2.5 h of extraction.

On NSCA all the laboratories had problems isolating putative *X. campestris* pv. *campestris* colonies in saprophyte-rich samples due to the overgrowth phenomenon. This was especially apparent after 2.5 h of extraction. Overgrowth of putative *X. campestris* pv. *campestris* colonies decreased in the following order: NSCA > NSCAA > mFS > CS20ABN. It should be noted that recovery of *X. campestris* pv. *campestris* on the four media is critical for their performance. Certain neomycin-sensitive *X. campestris* pv. *campestris* isolates might not be detected on CS20ABN (Schaad, unpublished data). Therefore the use of CS20ABN as a sole semi-selective medium is probably not advisable. The ability to identify putative *X. campestris* pv. *campestris* colonies on the four media was not equal for most laboratories and was dependent upon their experience with the different media. For example, using mFS several laboratories only selected saprophytic colonies for the pathogenicity assays rather than *X. campestris* pv. *campestris* colonies that were actually present. The phenomenon is probably due to the atypical appearance of *X. campestris* pv. *campestris* on this medium and might underestimate the usefulness of mFS. However, laboratories with experience with the mFS medium were well able to recognize *X. campestris* pv. *campestris* colonies on this medium. Both pathogenicity assays were able to discriminate between pathogenic *X. campestris* pv. *campestris* and non-pathogenic isolates including the control. Most laboratories had several escapes in case of the excised cotyledon assay. Again, it should be noted that most laboratories did not have any experience with this particular assay. In general the *X. campestris* pv. *campestris* symptoms were more distinct in the case of the cabbage plant assay.

There were also pitfalls in this first comparative test organized by ISHI. For instance, it appeared that a pilot comparative test should have been performed to check whether the protocol was clear and flawless. It also became very clear that the number of variables should be reduced in future tests. This will reduce the workload for participants and speed up the statistical analysis of the data by the organizers. Despite these problems all participants indicated that they had learned a lot from this test. A further development of the test is that several laboratories modified their procedures to detect *X. campestris* pv. *campestris*. For instance, several laboratories replaced the use of NSCA medium by another more selective medium.

ACKNOWLEDGEMENTS

I thank all participants in the USA, France, and The Netherlands for their hard work and give special thanks to the members of the International Technical Group Brassica Dick Morrison (USA), Hubert Lybeert (France), and Jeroen van Bilsen (The Netherlands) for their constructive help in the organization of this test.

REFERENCES

Chang, C.J., Donaldson, D., Crowley, M. and Pinow, D. (1991) A new semiselective medium for isolation of *Xanthomonas campestris* pv. *campestris* from crucifer seeds. *Phytopathology* 81, 449–453.

Randhawa, P.S. and Schaad, N.W. (1984) Selective isolation of *Xanthomonas campestris* pv. *campestris* from crucifer seeds. *Phytopathology* 74, 268–272.

Schaad, N.W. (1989) Detection of *Xanthomonas campestris* pv. *campestris* in crucifers. In: Saettler, A.W., Schaad, N.W. and Roth, D.A. (eds), *Detection of Bacteria in Seeds and Other Planting Material.* The American Phytopathological Society, St Paul, Minnesota, pp. 68–75.

Schaad, N.W. and Franken, A.A.J.M. (1996) *Xanthomonas campestris* pv. *campestris*. ISTA Handbook on Seed Health Testing. Working Sheet no. 50 (2nd edn).

Schaad, N.W. and Kendrick, R. (1975) A qualitative method for the detection of *Xanthomonas campestris* in crucifer seeds. *Phytopathology* 65, 1034–1036.

Wilson, E.E., Zeitoun, F.M. and Frederickson, D.L. (1967) Bacterial phloem canker, a new disease of Persian walnut trees. *Phytopathology* 57, 618–621.

Yuen, F.Y., Alvarez, A.M. Benedict, A.A. and Trotter, K.J. (1987) Use of monoclonal antibodies to monitor the dissemination of *Xanthomonas campestris* pv. *campestris. Phytopathology* 77, 366–370.

28

Comparative Tests with the Osmotic Blotter Method for Detection of *Drechslera* spp. in Barley Seeds

G. Brodal

Norwegian Agricultural Inspection Service, Seed Testing Station, Fellesbygget, N-1432, Ås, Norway

INTRODUCTION

Treatment of seeds with fungicides according to need requires rapid, simple and cheap seed health testing methods. The osmotic blotter method is a method widely used in Norway and Sweden for the detection of *Drechslera* spp. in cereal seeds (Joelsson, 1983; Brodal, 1991). Thousands of samples have been tested annually with this method to determine the need for seed treatment against seedborne *Drechslera teres* (Sacc.) Shoem., *D. graminea* (Rabenh. ex. Schlecht.) Shoem. in barley and *D. avenae* (Eidam) Scharif in oats. In spite of this routine use, very little information has been published on the sensitivity and reproducibility of this method. During the years 1990–1993, comparative tests with 57 barley seed samples using the osmotic blotter and the freezing blotter methods for detection of *Drechslera* spp. were carried out by a Nordic seed pathology working group (Brodal, 1995). The frequencies of *Drechslera* spp. recorded with the two methods were well correlated; however, the osmotic blotter method revealed about 40% higher frequencies of infection than the freezing blotter method. High correlations were also found between results with the osmotic blotter method from the six Nordic seed testing laboratories.

With the aim of gaining more experience with the osmotic blotter method, to evaluate the method more thoroughly, to test the reproducibility and to compare its sensitivity with other methods, an ISTA-PDC sub-Working Group on *Drechslera* spp. in barley was established in 1993. This paper presents the main results from a comparative test with the osmotic blotter, freezing blotter and agar plate methods carried out by 13 members of the working group, in

 Seed Health Testing (eds J.D. Hutchins and J.C. Reeves)

1994 and 1995 on ten barley seed samples. Four of the participating laboratories had a lot of experience with the osmotic blotter method, the others used the method for the first time.

MATERIALS AND METHODS

Ten barley seed samples, consisting of two common Norwegian cultivars, one two-rowed and one six-rowed type, known to be infected by *Drechslera* spp., were collected from seed lots harvested in Norway 1993 (Table 28.1). The *Drechslera* frequency in the selected samples ranged from 3 to 90%, recorded with the osmotic blotter method at the seed testing station in Norway.

In addition to the osmotic blotter test, ten of the laboratories carried out a freezing blotter test and six an agar plate test on the same seed samples (Table 28.2). With all methods, 400 seeds were tested per sample.

Osmotic Blotter Method

The method was, as far as possible, carried out by the same procedure used for routine testing in Norway and Sweden (Brodal *et al.*, 1994), which is a somewhat modified version of the method reported in Sweden (Joelsson, 1983; Svensson, 1983). Before plating, the seed samples were pretreated for 2 h at 90°C in open dishes, to reduce competition from saprophytic microorganisms. Seeds were placed on filter paper moistened with a sugar solution (0.5 M sucrose). The osmotic pressure inhibits the germination of the seeds. To keep the seeds in the correct position and to provide good contact between the seeds and the paper the seeds are plated into hollows, which are punched after soaking the paper. Only the laboratories using this method routinely had the equipment for this; others plated on flat paper. The samples were incubated in

Table 28.1. Barley seed samples used in comparative tests for *Drechslera* spp. (Tyra, two-rowed, and Bamse, six-rowed, are both Norwegian spring barley cultivars).

Sample number	Cultivar	% seed infected with *Drechslera* spp.
1	Tyra	69
2	Bamse	11
3	Bamse	43
4	Tyra	17
5	Bamse	85
6	Bamse	8
7	Tyra	3
8	Bamse	32
9	Bamse	15
10	Bamse	90

dishes with transparent covers for 7 days with alternating photoperiods: 16 h at 27 ± 1°C in white light, and 8 h at 21 ± 1°C in darkness in most laboratories. Since some laboratories do not use the method routinely they had difficulty achieving optimal incubation conditions. The described incubation conditions inhibit the development of conidia and mycelium, and favour development of brick-red pigment spots on the paper. *D. teres* and *D. graminea* are known to produce the anthraquinone catenarin (Engström *et al.*, 1993).

During examination, the seeds were removed, and a solution of 1% NaOH was added to the paper. The colour of the pigment spots then changed to pink–violet, which made the spots more visible. Recording was done by counting the spots by visual examination. The size of the spots usually varied from 1–2 mm up to 1–2 cm in diameter. When counting the spots it was an advantage to have the hollows in the paper, because it was then easy to see exactly where the seeds had been placed. This ensured the counting of the exact number of infected seeds, even if one seed has developed more spots, e.g. in both ends, and thereby being counted as two infected seeds. Some of the participants had difficulty in recognizing the correct colour of the pigment spots. Unfortunately, *D. graminea* and *D. teres* develop the same type of pigment, and therefore it is not possible to distinguish between them (Engström *et al.*, 1993). Distinguishing these two species is also a problem with other routine detection methods (Rennie and Tomlin, 1984).

Freezing Blotter and Agar Plate Tests

The methods were carried out as described in the Working sheet No. 6 in the ISTA *Handbook on Seed Health Testing* (Rennie and Tomlin, 1984).

Table 28.2. Particpating laboratories and the methods used to test barley seed samples infected with *Drechslera* spp. in 1994 and 1995.

Laboratory	Osmotic blotter	Freezing blotter	Agar plate
Ås, Norway	+	+	+
Beaucouze Cedex, France	+	–	+
Buenos Aires, Argentina	+	+	–
Cambridge, UK	+	–	+
Ede, The Netherlands	+	+	–
Edinburgh, UK	+	+	+
Hellerup, Denmark	+	+	–
Kvithamar, Norway	+	–	–
Lisbon, Portugal	+	+	+
Loimaa, Finland	+	+	+
Lyngby, Denmark	+	+	–
Ottawa, Canada	+	+	+
Svalöv, Sweden	+	+	–

+, Test conducted; –, no test conducted.

RESULTS AND DISCUSSION

The frequencies of infection of *Drechslera* spp. in each laboratory, as an average for the ten seed samples, recorded with the osmotic blotter, the freezing blotter and the agar plate methods are shown in Fig. 28.1. With the osmotic blotter method, some laboratories were more successful in recording infections than others. Results from the most experienced laboratories were in the range of 35–37% on average for the ten samples. The inexperienced laboratories recorded an average seed infection of 11–32% with the osmotic blotter method. Despite the difference in detection levels, the infection frequencies recorded for the series of individual samples were in most cases well correlated between the 13 laboratories (Table 28.3). Of 78 correlations, 51 achieved a correlation coefficient (r) of > 0.90 and only nine had $r < 0.80$. It is believed that both suboptimal incubation conditions as well as lack of experience in recording pigment spots contributed to these systematic differences in detection levels.

The infection frequencies recorded with the freezing blotter method in the different laboratories varied between 13% and 24% on average for the ten samples (Fig. 28.1). Infection frequencies recorded for the series of individual samples were well correlated between the ten laboratories. Of 45 correlations, 33 had an r of > 0.90 and 1 had an r of < 0.80 (data not shown). The infection frequencies recorded with the agar plate method varied between 16% and 39%

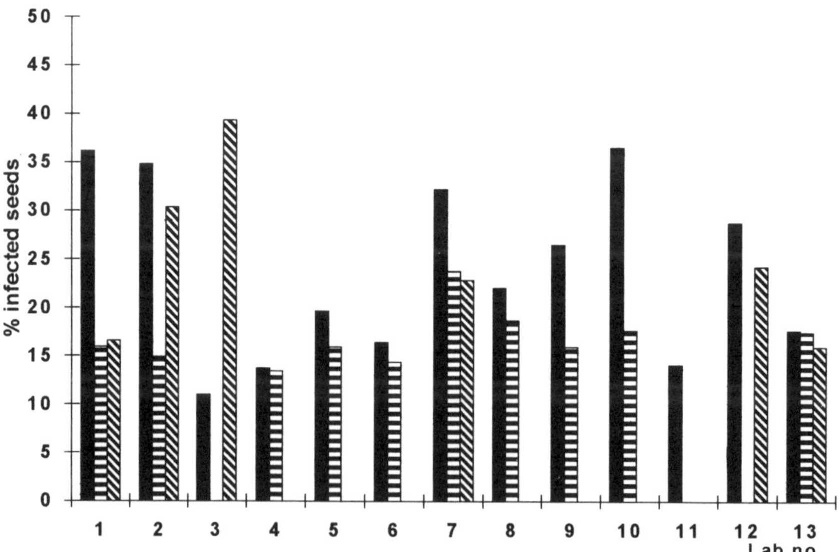

Fig. 28.1. Average frequency of infection with *Drechslera* spp. for ten barley seed samples, recorded with the osmotic blotter, freezing blotter and agar plate tests in 13 laboratories. ■ Osmotic blotter; ⬓ freezing blotter; ▧ agar plate.

on average for the ten samples (Fig. 28.1). Infection frequencies recorded for the series of individual samples in the six laboratories were correlated too (data not shown), except for the results from one laboratory with little experience of the method. The high average infection frequencies recorded with the agar plate method in laboratories 2 and 3 were caused by results obtained from two samples that differed significantly from results recorded in the other laboratories.

A high level of agreement was found between individual results with the osmotic blotter method in four laboratories that were experienced, or considered to be the most successful with the method (Table 28.4). The average

Table 28.3. Correlation coefficient (*r*) between frequencies of *Drechslera* spp. in barley seeds in 13 laboratories recorded with the osmotic blotter method.

Laboratory no.	1	2	3	4	5	6	7	8	9	10	11	12
2	0.98	0.86										
3	0.87	0.93	0.90									
4	0.97	0.68	0.55	0.80								
5	0.77	0.99	0.86	0.93	0.63							
6	0.97	0.92	0.88	0.99	0.82	0.90						
7	0.96	0.95	0.95	0.96	0.67	0.96	0.94					
8	0.97	0.95	0.93	0.96	0.70	0.96	0.94	1.00				
9	0.97	0.93	0.87	0.96	0.80	0.93	0.94	0.96	0.97			
10	0.98	0.88	0.73	0.92	0.89	0.84	0.90	0.84	0.84	0.88		
11	0.92	0.95	0.80	0.95	0.84	0.94	0.94	0.93	0.95	0.96	0.95	
12	0.98	0.89	0.95	0.94	0.70	0.89	0.94	0.98	0.98	0.96	0.78	0.90

Table 28.4. Percentage of seeds infected with *Drechslera* spp. as assessed by the osmotic blotter method in four laboratories.

	Laboratory number			
Sample no.	1	2	3	4
1	67	68	63	65
2	9	10	9	10
3	39	45	34	37
4	18	21	19	18
5	83	82	51	84
6	7	7	9	7
7	5	1	3	2
8	30	10	37	46
9	12	13	16	15
10	92	91	83	81
Average	36	35	32	37

results obtained with the osmotic blotter method from these laboratories were used for comparisons with the other methods. A linear relationship between frequencies of *Drechslera* infection recorded by the osmotic blotter and the freezing blotter method are shown in Fig. 28.2. Approximately 50% higher infection frequencies were recorded with the osmotic blotter method than with the freezing blotter method. This is in agreement with a previous comparative test carried out in the Nordic countries (Brodal, 1995). A linear regression was also found between results with the osmotic blotter and the agar plate method (Fig. 28.3). The osmotic blotter method revealed about 20% more *Drechslera* than the agar plate method. The relationship between the results with these two methods is shown in Fig. 28.4. The freezing blotter method detected only about 70% of the *Drechslera* recorded with the agar plate method.

In this study, the time needed to carry out the different methods was not measured. However, the osmotic blotter method was considered to be less time consuming than the other methods. The main reason for this was the quick examination based on pigment production. Also, when plating seeds with a vacuum counter, without the need for sterile conditions, plating 50 or 100 seeds per dish contributes to the efficiency of the method. In well organized routine testing laboratories, one analyst can easily handle 35–40 samples in one day (200 seeds per sample), including pretreatment, preparation of dishes, plating, moving in and out of incubator, examination, recording and reporting of results.

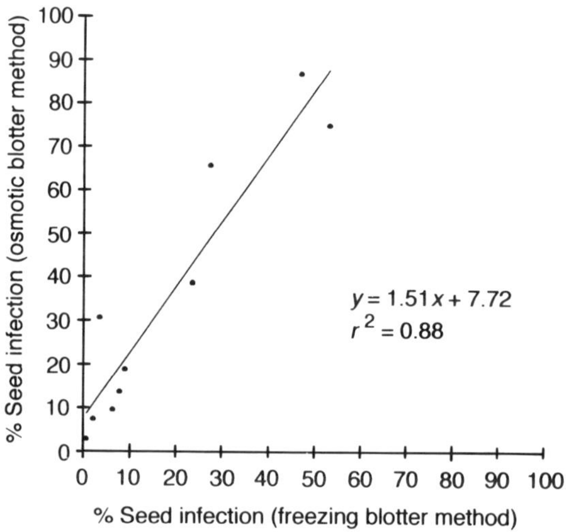

Fig. 28.2. Relationship between frequencies of infection by *Drechslera* spp. recorded by the osmotic blotter and the freezing blotter methods.

CONCLUSIONS

The osmotic blotter method gave reproducible results for *Drechslera* spp. in barley seeds when used by experienced laboratories. To obtain reliable results in inexperienced laboratories training with this method is required and incubation conditions need to be standardized. In this study, the osmotic blotter

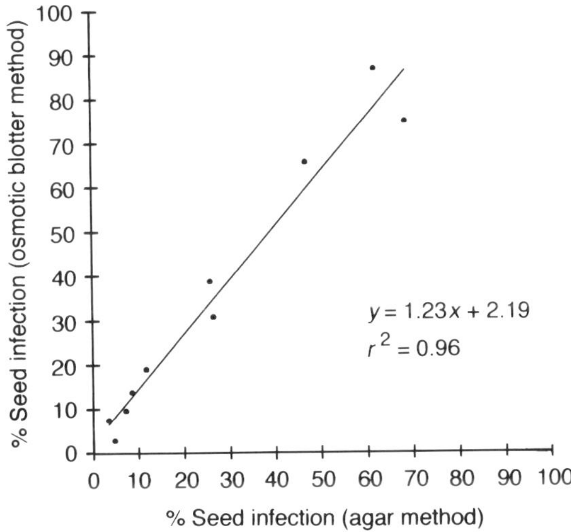

Fig. 28.3. Relationship between frequencies of infection by *Drechslera* spp. recorded by the osmotic blotter and the agar methods.

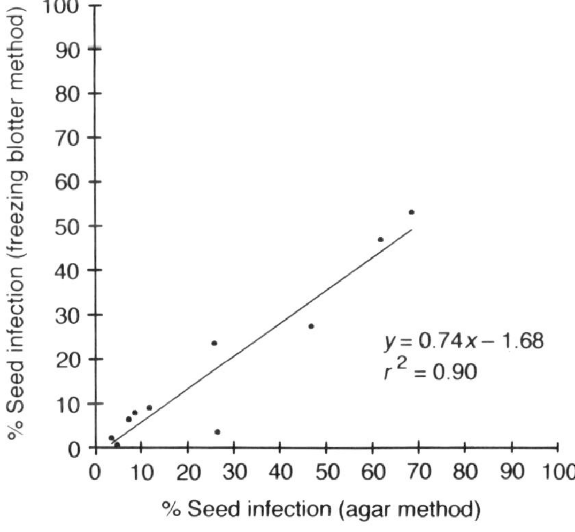

Fig. 28.4. Relationship between frequencies of infection by *Drechslera* spp. recorded by the freezing blotter and the agar methods.

method was more sensitive than the freezing blotter and the agar plate methods. The method was both quick and easy to carry out, with no need for a microscope, preparation of media or sterile conditions and is therefore well suited for routine use. Based on the results from the experienced laboratories in this study and from a previous study in the Nordic countries (Brodal, 1995), the osmotic blotter method is recommended when there is a need for testing of a large number of samples at a low cost, as in routine testing for assessing the need for seed treatment against *Drechslera* spp. in barley.

ACKNOWLEDGEMENTS

I want to thank Signe Dahl for her assistance with statistical analyses of data and preparation of tables and figures. The following members of the ISTA-PDC sub-Working Group on *Drechslera* spp. in barley participated in the test reported in this paper: D. Barreto and M. Carmona (Argentina), R.J.H. van den Born (The Netherlands), G. Brodal (Norway), A.M. Halkilahti (Finland), A.L. Hynkemejer (Denmark), M. Moe (Norway), V. Molinero-Demilly (France), M. Malheiro (Portugal), V. Cockerell and J. Reeves (UK), J. Sheppard (Canada), C. Scheel (Denmark) and K. Sperlingsson (Sweden).

REFERENCES

Brodal, G. (1991) Såkornbeising etter behov. Nye analyserutiner ved Statens frøkontroll frahøsten 1990 [Fungicide treatment of cereal seed according to need. New routine analyses at the State Seed Testing Station from autumn 1990]. Statens Fagtjeneste for landbruket. *Faginfo* 2, 205–211.

Brodal, G. (1995) The osmotic method for detection of *Drechslera graminea/D. teres* on barley seed as compared with the freezing blotter method. In: *Abstracts 24th ISTA Congress Seed Symposium, Copenhagen, Denmark, June 7–16, 1995*, p. 24.

Brodal, G., Halkilahti, A.M., Jørgensen, J. and Sperlingsson, K. (1994) Frøbårne plantesjukdommer og frøpatologiske undersøkelser [Seed borne plant diseases and seed pathological investigations]. *TemaNord* 630, Nordisk Ministerråd, Copenhagen, 50 pp.

Engström, K., Brishammar, S., Svensson, C., Bengtsson, M. and Andersson, R. (1993) Anthraquinones from some *Drechslera* species and *Bipolaris sorokiniana*. *Mycological Research* 97, 381–384.

Joelsson, G. (1983) The osmotic method. A method for rapid determination of seed-borne fungi. Preprint no. 104, 20th ISTA Congress, Ottawa, Canada. 9 pp.

Rennie, W.J. and Tomlin, M.M. (1984) Barley leaf stripe. Working Sheet no. 6 (2nd edn). In: *ISTA Handbook on Seed Health Testing*, Section 2: Working Sheets. International Seed Testing Association, Zurich, 4 pp.

Svensson, C. (1983) Osmotic method for detecting *Drechslera* spp. on barley seed. *Abstracts of Papers of the 4th International Congress of Plant Pathology*, p. 399.

ISTA Accreditation in Seed Health Testing Laboratories

29

J.D. Hutchins

National Institute of Agricultural Botany, Huntingdon Road, Cambridge CB3 OLE, UK

INTRODUCTION

Seeds are an internationally important commodity, the quantity and value of seed traded is high, and approximately 90% of all food crops are propagated from seed (Schwinn, 1994). The World Trade Organization, formerly the General Agreement on Tariffs and Trade (GATT) ensures that seed trade is facilitated through the removal, or reduction, in barriers to trade. The maintainance of seed quality (freedom from contamination and disease) is dependent upon accurate sampling procedures and laboratory testing. If the procedures used are sufficiently robust, then the results obtained from testing representative samples may be used to infer the quality of the seed lot (ISTA, 1993) and hence to compare the value of different lots. The International Seed Testing Association (ISTA) only issues International Certificates to seed lots that have been tested in an ISTA laboratory (Wold, 1996). Issuance of the certificate allows the seed to be traded. Acceptance of the quality implied by the certificate is not, however, universally accepted by other organizations or countries (S. Cooper, personal communication).

International acceptance of test results will only be achieved through standardization of sampling and testing methods which is ISTA's principle aim (ISTA, 1993). Such methods are now more commonly integrated into formal quality assurance schemes. The ISTA has defined an accreditation standard, based upon worldwide accepted standards, to be implemented in seed testing laboratories. The reliability, accuracy and repeatability of data from such accredited laboratories should result in mutual acceptance of test results. The aim of this paper is to provide an overview of the ISTA accreditation standard and to indicate, where possible, the ISTA Plant Disease Committee (ISTA-PDC) has a role to play in developing the standard in seed health testing laboratories.

ISTA ACCREDITATION STANDARD

The accreditation standard, covering all aspects of seed sampling and laboratory testing, has been prepared to meet the specific needs of ISTA, its member laboratories and the international seed trade. The standard is based upon ISO Guide 25 and EN 45001 standards and was approved by the ISTA Executive Committee (ISTA, 1996). These standards have been used as a guide for similar accreditation standards (J. Sheppard and H. Wesseling, personal communication). To achieve the status of an accredited ISTA laboratory, each laboratory must produce a Quality Manual which is submitted to the ISTA Secretariat. The applicant laboratory will be expected to meet predefined organizational standards and show competence in both a seed testing programme and laboratory inspection. ISTA accreditation will be granted if the standard is fully met and the appropriate fees paid. The Quality Manual describes the quality system operated in the laboratory (Table 29.1), defining how the laboratory functions and what measures are taken to ensure that pre-defined standards are adhered to.

General Policies

The general policies described in the Quality Manual outline the function and operation of the laboratory and must clearly state that the service offered will be in accordance with the procedures outlined in the ISTA Rules. The process of auditing and reviewing the quality system must be described and conducted on an annual basis. The reviews should be recorded, together with the details of any corrective actions taken. Quality control policies specific to the identification of seed lots, sampling arrangements and laboratory testing procedures

Table 29.1. Selected components of the quality system that should be described in the laboratory Quality Manual. Summarized from the ISTA accreditation standard (ISTA, 1996).

Section	Outline
General policies	statement and function of services; quality control checks; inter-laboratory comparisons; audit and review
Organization of laboratory	staff responsibilities; staff training
Methods and procedures	test procedures; support services; administrative procedures; monitoring programmes
Calibration	calibration of equipment; reference materials
Records	seed testing records; system audits; equipment maintenance; calibration records

must be described. The checking of seed health testing procedures should include methods to verify data obtained from tests and arrangements for the monitoring of analyst performance. The general arrangements for inclusion in inter-laboratory comparisons should be defined and used to monitor the performance of the laboratory by comparison with similarly accredited laboratories. These tests should be conducted according to defined procedures.

Organization of Laboratory

The structure and organization of the laboratory should be defined in the Quality Manual and must include the responsibilities of individual members of staff. Laboratory staff should have and maintain the necessary education, training, technical knowledge, demonstrated skills and experience for their assigned functions.

Methods and Procedures

The methods and procedures used for seed sampling and testing should be adequately documented and available to all members of staff and must be sufficiently detailed to allow them to be followed, both accurately and reproducibly. These procedures must cover all aspects of the work involved, including test methodology and any related support services. The latter may include preparation of agar media or stock solutions of chemicals and buffers. Procedures describing the use of items of equipment (including balances, microwave ovens, drying ovens, autoclaves and pH meters) and the monitoring, repair and servicing of these items must also be written. The administrative procedures used in the laboratory should be similarly described including the system used for identifying samples and test results. Procedures defining the control of laboratory documentation, staff training programmes, arrangements for dealing with complaints, external customers or suppliers and any monitoring programmes conducted in the laboratory must also be written. For example, in many agar plate tests the temperature or lighting intervals used may be of critical importance (Neergaard, 1977) and should be monitored to ensure samples are tested under the appropriate conditions.

Calibration and Testing Materials

Sampling, measuring and testing equipment shall be calibrated where appropriate before being put into service and thereafter according to an established programme. The overall calibration programme must be designed to ensure that all measurements made in the laboratory are traceable to national/ international standards of measurement through an unbroken chain of traceability. Appropriate samples and standards should be used, including standard buffer solutions for pH meters, calibration weights for balances and reference materials for seed health tests. The quality of testing materials (agar media, blotters, water and chemicals, etc.) used in seed health tests should be checked periodically.

Records

The laboratory must define and maintain a record system to suit its own circumstances. Records should usually be kept for a period of at least 3 years and contain all original observations, calculations and derived data relating to the conduct of an individual test. This will also include: (i) an up-to-date record of laboratory staff and their training; (ii) records relating to results, calibration, maintenance and repair of equipment and reference materials; and (iii) all audit reports, records of management reviews, test reports and certificates.

DISCUSSION

The ISTA-PDC has been involved in the development of standardized test procedures through a programme of research and comparative tests (Neergaard, 1970; Hewett, 1989; Langerak, 1992). For many years the PDC has aimed at standardizing and improving seed health testing methods, especially for fungi and bacteria (Franken and Sheppard, 1993). The PDC should therefore have an active role to play in the development of ISTA accreditation in seed health testing laboratories. The main thrust of activity required is shown in Table 29.2.

Methodology

Methods used in routine seed health testing laboratories vary quite widely (Maude, 1995) and are described in the ISTA *Handbook on Seed Health Testing*. The Working Sheets, each dealing with one pathogen on one host, are used by many laboratories as their main source of information. However, they are often not sufficiently accurate or detailed (Franken and Sheppard, 1993). For example, the Working Sheet on *Tilletia caries* on wheat (Kietreiber, 1984) does not provide sufficient information regarding sample preparation in routine testing laboratories (V. Cockerell, personal communication). Although guidance exists for conduct of seed health tests, little information is available for preparation of supplementary reagents including agar media and chemical solutions which have been reported to influence test results (Neergaard, 1970;

Table 29.2. Main requirements of the ISTA-PDC necessary for the development and implementation of the ISTA accreditation standard.

Area	Requirements
Methods	standardization and definition of methods
Tolerances	definition of acceptable tolerances for variation in seed health tests
Quality control	QC monitoring of individual analyst performance
Inter-laboratory comparisons	QC monitoring of laboratory performance
Reference materials	reference materials for use in seed health tests

Franken and Sheppard, 1993). The quality of agar media is of paramount importance. In routine clinical laboratories, the quality of each batch of agar media is monitored by inoculation with reference organisms (Mossel *et al.*, 1983).

Developments in seed health testing methodology are regularly published in the scientific literature, and include enzyme-linked immunosorbent assay (ELISA) (van Vuurde and Maat, 1983) and polymerase chain reaction (PCR) based techniques (Reeves, 1995). New techniques must be readily adapted for use in routine testing laboratories and should be evaluated, compared and standardized using appropriate guidelines, which do not currently exist. Langerak and Franken (1995) describe a draft procedure for the conduct of collaborative studies, based on an internationally accepted protocol (Anon., 1989). Once evaluated these new methods may be integrated into the laboratory QA system.

Tolerances

Variation in the results of replicate sub-samples of a seed test may be due to sampling variation, differences in test methodology and differences in assessment. The ISTA Rules contain guidelines for acceptable levels of variation between replicate working samples, from the same submitted sample, for both purity and germination tests (Miles, 1963). No such guidelines exist for seed health tests (Mangan, 1983). There have been limited attempts to address this discrepancy: Rennie and Tomlin (1984) and Jorgensen (1982, 1983) used Miles' tables to evaluate their data. They concluded that these tables could be used to measure replicate variation in tests performed with infected cereal seed. Similarly, Hewett (1983, 1987a) used tolerance tables to characterize sample variation in both barley loose smut and *Ascochyta pisi* tests. The PDC must develop this work further. Without these techniques, acceptable variations in seed health data cannot be accurately quantified.

Quality Control of Analyst Performance

Appropriate quality control procedures may be used, within a laboratory, to monitor individual analyst performance aiming to reduce operator variation. Data obtained from identical samples can be compared using defined tolerance tables. The technique of re-examination of the same sample (Hewett, 1983) could be extended to other tests, e.g. re-examination of filters used in the bunt test. If the seed sample data were outside of tolerance the sample may be retested or the analyst retrained. Sources of analyst variation can be avoided by careful and comprehensive training and corrected, once identified, by subsequent refresher periods (Hewett, 1983). The PDC has a role to play in developing rigorous quality control programmes. To date tolerance tables have been applied to traditional seed health testing methods for fungal pathogens. There is an increasing need to apply these techniques to both bacterial and viral pathogens. In the future, these methods will need to be applied to the results obtained from new seed health testing technologies.

Comparative Trials

Routine inter-laboratory tests, with the aim of monitoring laboratory performance, form an integral component of the quality system of accredited laboratories. Such tests are commonplace in microbiology laboratories testing foodstuffs and water. Comparative trials (e.g. Hewett, 1987a; Rennie and Tomlin, 1984) with the aim of evaluating new test procedures, have been the backbone of the ISTA-PDC's work (Neergaard, 1970; Hewett, 1989). As part of the ISTA accreditation standard, inter-laboratory comparisons as a method of monitoring laboratory performance will need to be conducted. Although ISTA have a defined schedule for comparative testing (referee testing) of purity and germination tests (Wold, 1996), the PDC has no defined protocol for such tests. In order to develop such programmes, adequate guidelines should be produced. Using the draft procedure described by Langerak and Franken (1995), such guidelines could be adapted for routine inter-laboratory tests. These guidelines should be drawn up based on appropriate international standards and in consultation with the ISTA Referee Testing Committee.

Reference Materials

As part of an accurate testing procedure calibrated reference materials are used to validate the authenticity of tests performed on uncharacterized test samples. Biological reference materials are scarce but have become increasingly important in all aspects of microbiological testing. For seed health tests, the most logical reference material would be a seed lot, infected with a known quantity of pathogen inoculum. Such reference materials could be obtained either naturally or artificially (by mixing known infected seed with uninfected seed) and stored to retain its characteristics. Without further investigation, infected seed stocks cannot be relied upon to be used as reference materials because of the variability in pathogen recovery (Taylor *et al.*, 1979; Hewett, 1987b). Reference materials have been developed for use in food and water testing (In't Veld and Notermans, 1992). Similar materials could theoretically be developed for seed health testing. Franken and Sheppard (1993) suggested that the PDC initiated research in this area. The development of bacterial reference materials for use in seed health testing has now started (see Chapter 31). Reference materials for other priority pathogens will similarly need developing.

CONCLUSION

This paper has described the ISTA accreditation standard and the process of obtaining laboratory accreditation. As part of this process, the seed health testing laboratory will be presented with specific problems: method standardization, assessment of variation in seed health tests, quality control in the laboratory, inter-laboratory testing and the need for reference materials. The ISTA-PDC has a coordinating role to play in the initiation, development and production of guidelines and procedures to meet these demands.

References

Anonymous (1989) Guidelines for collaborative study procedures. *Journal of the Association of Official Analytical Chemists* 72, 694–704.

Franken, A.A.J.M. and Sheppard, J.W. (1993) The need for accurate guidelines in comparative testing between seed-health testing laboratories. In: Sheppard, J.W. (ed.), *Proceedings of First ISTA Plant Disease Committee Symposium, Ottawa, Canada*, pp. 143–148.

Hewett, P.D. (1983) Assessing agreement in the embryo test for loose smut of barley. *Journal of the National Institute of Agricultural Botany* 16, 279–284.

Hewett, P.D. (1987a) Detection of seed-borne *Ascochyta pisi* Lib. and test agreement within and between laboratories. *Seed Science and Technology* 15, 271–283.

Hewett, P.D. (1987b) Pathogen viability on seed in deep freeze storage. *Seed Science and Technology* 15, 73–77.

Hewett, P.D. (1989) Report of the Plant Disease Committee 1986–89. *Seed Science and Technology*, 17 (Suppl. 1), 97–102.

In't Veld, P.H. and Notermans, S. (1992) Use of reference materials (spray-dried milk artificially contaminated with *Salmonella typhimurium*) to validate detection methods for *Salmonella*. *Journal of Food Protection* 55, 855–858.

ISTA (1993) International rules for seed testing. *Seed Science and Technology*, 21 (Suppl.), 288 pp.

ISTA (1996) *ISTA Seed Testing Laboratory Accreditation Standard.* ISTA, Zurich, 10 pp.

Jorgensen, J. (1982) The freezing blotter method in testing barley seed for inoculum of *Pyrenophora graminea* and *P. teres*. Repeatability of test results. *Seed Science and Technology* 10, 639–646.

Jorgensen, J. (1983) Disease testing of barley seed and application of test results in Denmark. *Seed Science and Technology* 11, 615–624.

Kietreiber, M. (1984) Bunt (stinking smut). In: *Handbook on Seed Health Testing*, Working Sheet no. 53, ISTA, Zurich.

Langerak, C.J. (1992) Report of the Plant Disease Committee. *Seed Science and Technology*, 20 (Suppl. 1), 115–122.

Langerak, C.J. and Franken, A.A.J.M. (1995) Diagnostic methods for the detection of plant pathogens in vegetable seeds. In: Martin, T. (ed.), *Seed Treatment: Progress and Prospects*. Monograph no. 57, BCPC, pp. 169–178.

Mangan, A. (1983) The use of plain water agar for the detection of *Phoma betae* on beet seed. *Seed Science and Technology* 11, 607–614.

Maude, R. (1995) *Seedborne Diseases and their Control. Principles and Practice*. CAB International, Wallingford, 280 pp.

Miles, S.R. (1963) Handbook of tolerances and of measures of precision for seed testing. *Proceedings of the International Seed Testing Association* 28, 525–686.

Mossel, D.A.A., Bonants-Van Laarhoven, T.M.G., Ligtenberg-Merkus, A.M.T. and Werdler, M.E.B. (1983) Quality assurance of selective culture media for bacteria, moulds, and yeasts: an attempt at standardisation on the international level. *Journal of Applied Bacteriology* 54, 313–327.

Neergaard, P. (1970) Seed pathology, international co-operation and organisation. *Proceedings of the International Seed Testing Association* 35, 19–42.

Neergaard, P. (1977) *Seed Pathology*. MacMillan Press, London, 839 pp.

Reeves, J.C. (1995) Nucleic acid techniques in testing for seedborne diseases. In: Skerritt, J.H. and Appels, R. (eds), *New Diagnostics in Crop Sciences*. CAB International, Wallingford, pp. 127–149.

Rennie, W.J. and Tomlin, M.M. (1984) Repeatability, reproducibility and interrelationship of tests on wheat seed samples infected with *Septoria nodorum*. *Seed Science and Technology* 34, 417–428.

Schwinn, F.J. (1994) Seed treatment – a panacea for plant protection? In: Martin, T. (ed.), *Seed Treatment: Progress and Prospects*, Monograph no. 57, BCPC, pp. 3–14.

Taylor, J.D., Dudley, C.L. and Presley, L. (1979) Studies of halo-blight seed infection and disease transmission in dwarf beans. *Annals of Applied Biology* 93, 267–277.

van Vuurde, J.W.L. and Maat, D.Z. (1983) Routine application of ELISA for the detection of lettuce mosaic virus in lettuce seeds. *Seed Science and Technology* 11, 505–513.

Wold, A. (1996). ISTA – history 1974-1995. *Seed Science and Technology* 24, 95–106.

30

NAL: a Quality Assurance System for Seed Health Testing Laboratories

J.B.M. Wesseling

NAKG, Sotaweg 22, 2371 GD Roelofarendsveen, The Netherlands

INTRODUCTION

Seed companies spend a lot of energy in seed health tests and have the knowledge and equipment to perform these tests. However, the results of tests conducted by companies cannot be used as official quality and health certificates because company tests are not considered to be independent. This approach has several disadvantages for the companies and their customers: it is expensive, involves retesting (often against different standards, with all the risks concerned), is time consuming and has extra logistic problems: an official test often takes at least 3 weeks.

In line with the system approach (monitoring of companies) in the inspection of vegetable and flower seeds in The Netherlands, the NAKG has developed an accreditation system for seed testing laboratories to overcome this problem. The system is designed to guarantee that tests are performed correctly, reliably and reproducibly, that whichever method is used the results obtained are accurate and comparable and that the customer or any other interested person or institute can rely upon them. Accredited companies should then be able to use their certificates for customers instead of official certificates.

THE STRUCTURE OF NAL

The NAKG accredited laboratory (NAL) accreditation system consist of four different elements: (i) general requirements for the organization of the

 Seed Health Testing (eds J.D. Hutchins and J.C. Reeves)

laboratory and supporting processes; (ii) specific requirements for the test methods which are used for the different seedborne pathogens; (iii) internal reference tests for the tests performed by the laboratory; and (iv) inter-laboratory tests (comparative tests) to check the performance of the laboratory on a regular basis. The requirements are based on EN-45001 (CEN/CENELEC, 1989), ISO 9001 (International Standard Organization, 1994), ISO Guide 25 (International Standard Organization, 1990) and NAKG experiences of the last 7 years in operating accreditation systems. The structure of the NAL system is outlined in Fig. 30.1. The different parts of the NAL system are described below.

GENERAL REQUIREMENTS

To perform seed health tests it is necessary that the laboratory has a full control over its organization and processes. Without such control, variability in the performance of the laboratory will be too large and will lead to too many

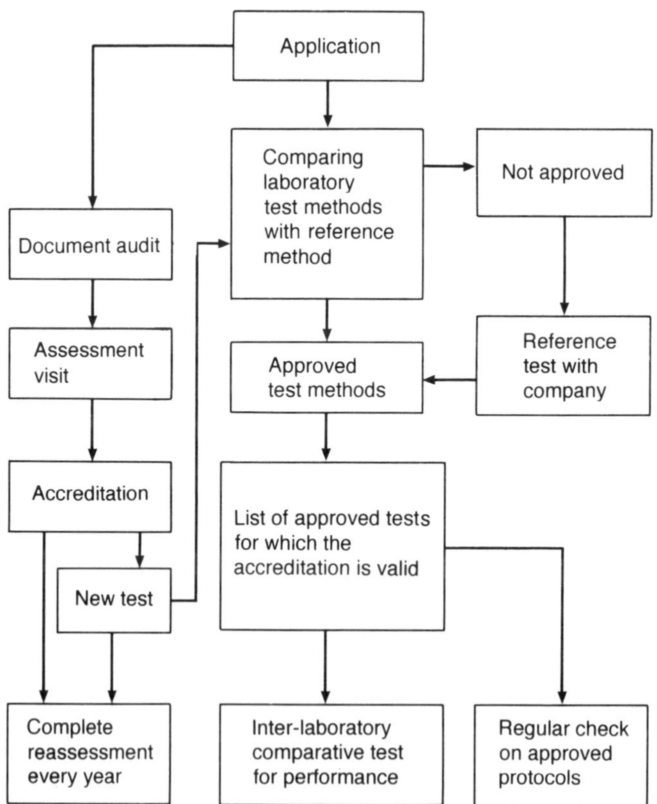

Fig. 30.1. The structure of the NAL accreditation system.

mistakes being made, which will have nothing to do with the testing itself. Only with a documented quality management system is it possible to achieve full control and create the framework with which to maintain and improve the system. The elements of this quality management system are listed in Table 30.1.

It is beyond the scope of this article to explain the different elements and their function in a quality management system. For the laboratory it is very important that staff have a thorough knowledge of the requirements and that they are able to implement them in harmony with their organization. The development and implementation of quality management systems often fails (in up to half of the first attempts) because the quality management system is not adapted to the organization. One of the main reasons is that many organizations copy the system from other organizations or models without adapting them for their own needs. Another important reason is that personnel are not adequately involved in the development and implementation of the quality management system.

The general elements listed in Table 30.1 are fundamental to the NAL system. When a laboratory meets these requirements it has a stable, reliable and controlled organization. Independent recognition of this is given by NAL accreditation after the NAKG have completed a documentation audit and a laboratory assessment (see Fig. 30.1).

TEST METHOD

An accurate test protocol is required to perform successfully a reliable test to detect a seedborne pathogen. To check this it is necessary to have reliable and

Table 30.1. General elements of the quality management system.

Policy of laboratory
Quality system and quality plan
Organization and administration
Personnel and training
Audit system (internal evaluation)
Document system
Analyzing of equipment
Test methods and procedures
Calibration and reference materials
Environment
Corrective and preventive measures
Analysis files and test reports (company certificates)
Sub-contracting test to third parties
External supporting services and provisions
Complaints

internationally accepted reference methods, which act as a reference with which seed company protocols can be compared and evaluated. In the approval of the company protocol the following points are investigated: (i) the description of the method (protocol); (ii) the chance of false negatives (should be less than 5%); (iii) the sensitivity and specificity of the method; (iv) the number of seeds tested per subsample and the quantity of seed to be tested; and (v) the suitability of the protocol in relation to the amount of seeds which have to be tested (can the protocol be used reliably in practice?).

Figure 30.1 shows how the check on the company protocol is carried out. First, the company protocol is compared with the reference protocol on paper and judged on the above-mentioned criteria. When requirements for all five points are satisfied then the method is approved. When there is any doubt or the company protocol differs greatly from the reference method, the approval has to be based on a reference test, with a positive result between the NAKG and the company.

This approach permits maximum flexibility concerning the methods that can be used by a laboratory, without decreasing the reliability of the tests that are performed. Company laboratories are therefore not restricted to one specific method and can develop their own according to their business requirements. There is as yet no clearly identified reliable reference method for many seed-borne pathogen tests. For germination, purity, moisture etc. the ISTA have produced an infrastructure, which recognizes a standard reference method. For seed health testing the situation is much worse. This is why the NAKG strongly supports the International Seed Health Initiative (see Chapter 12).

INTERNAL REFERENCE TEST

In seed health laboratories a lot of work is carried out by skilled personnel and involves much observational work. Many judgements are made by eye and hence involve a degree of subjectivity. However, when conducting a test the variation between personnel should be as small as possible; therefore internal reference tests are necessary to monitor analyst performance. There are two systems for reference testing: (i) retesting of samples or reference standards (performed blind, so that the identity of the sample is unknown) and (ii) examination of the same sample by all personnel who are allowed to perform the evaluation of that test. Both methods are important and should, wherever possible, be applied. In most laboratories the first method is used for internal reference. On the basis of statistical analysis it is determined whether the analyst performance is within the prescribed limits. The disadvantage of this approach is that the variability in seed health test results can be high and therefore the usefulness of this approach can be questioned. Associated with this is the problem that a stable reference standard is not available, which means that the variation within the retested sample has to be included in the analysis of the results. When it concerns examination by eye this variation is too high. Calibration of visual examination used in a test requires much lower limits of variation. The second method for internal reference test fulfils this demand. In fact the variation in examination should be zero. Every person examines the same sample and should reach the same conclusion. Directly after running the test, the results need to be compared and discussed. This method of internal reference is not frequently applied but for the calibration of subjective skills it is very important.

In the NAL system the internal reference tests of the laboratory are investigated in every assessment. The method of internal reference is examined in relation to type of test and special attention is given to the calibration of examinations conducted by eye. The results of the internal reference test are evaluated every reassessment (see Fig. 30.1) and are important in deciding whether to maintain the status of accreditation.

INTER-LABORATORY TEST

An important factor in a laboratory accreditation system, especially when results are used for external purposes, is the overall performance of a laboratory in relation to other laboratories that are performing the same tests. Interlaboratory tests are used to compare laboratory performance in an independent manner and are the only tool with which to assess whether the results of accredited laboratories performing the same test are in line with each other. It is also a very important check on the performance of the NAL system itself.

Laboratories who have the NAL status are obliged to participate in the inter-laboratory test programme of the NAKG. This programme depends on the number of tests the laboratory is accredited to perform. Every year they have to participate in at least one inter-laboratory test. Each accredited test must be checked at least every 3 years. The results of the inter-laboratory tests influence the decision as to whether a test will stay on the list of accredited test methods for the laboratory and whether the accreditation status can be maintained (see Fig. 30.1).

RESULTS

Four laboratories have so far been accredited using this system, and another four have applied for accreditation. The NAKG also runs another accreditation system, the Seedborne Pathogen Equivalence System, which concentrates on the test method and only components of the general elements of the NAL system. Experience gained from this system has revealed that most problems are caused by a lack of control over organization and laboratory processes (e.g.

insufficient arrangements for calibration, insufficient maintenance structure for reference material, no arrangements for availability of replacement items of equipment, differences in interpretation of results between personnel, unclear responsibility for the control of test results and undocumented changes in test protocols). These problems all indirectly lead to the removal of test methods from the list of approved test methods, or in one case, the loss of equivalence for all their approved test methods for 1 year. This system will therefore be replaced by the NAL system which addresses the issues of organization control.

The advantages of a complete accreditation system for companies are that they can use their own test results for official purposes. This prevents retesting by official institutes or can replace official testing for the customer. This saves time, money and logistic problems due to all tests taking less time. New tests can be introduced more easily and on a broader scale. The laboratory, its organization and its internal quality system are checked every year by specialists, keeping the laboratory up to date, preventing errors and mistakes, and reducing the costs of failing quality.

Most importantly, this kind of management adds value to both the seed company and its customers. This is demonstrated by the fact that in the past 3 years more company certificates have been accepted for official purposes by their customers. An additional advantage is that both the companies and their customers have become more aware of their role in defining and structuring the quality of seeds.

CONCLUSION

An accreditation system is an efficient way of controlling quality. Each company has a major economic stake in ensuring the quality of its products and, therefore, should be trusted to take responsibility for its own quality control. The system avoids duplication of effort through retesting by inspection services because it is no longer required. Finally, new quality control elements can be incorporated very quickly into the NAL system.

REFERENCES

CEN/CENELEC (1989) *General Criteria for the Operation of Testing Laboratories*. EN/45001. CEN/CENELEC, Brussels.

International Standard Organization (1990) *General Requirements for the Competence of Calibration and Testing Laboratories*. ISO Guide 25, third edition. ISO, Geneva.

International Standard Organization (1994) *ISO 9001, Model for the Quality Assurance in Design, Development, Production, Installation and Servicing*. ISO, Geneva.

31 Microbiological Reference Materials for Quality Control in Seed Health Testing

R.W. Van den Bulk[1] and J.D. Taylor[2]

[1]DLO-Centre for Plant Breeding and Reproduction Research (CPRO-DLO), Wageningen, The Netherlands; [2]Horticulture Research International (HRI), Wellesbourne, Warwick CV35 9EF, UK

INTRODUCTION

The use of disease-free planting material is a necessary component of an economically profitable and environmentally safe crop production. Healthy seed is the first prerequisite for obtaining a high quality end-product. Production and selection of healthy seed can be ensured by testing seed for the presence of unwanted microorganisms, such as plant pathogenic bacteria, as indicated in national and international plant health regulations. The International Seed Testing Association (ISTA) and the European and Mediterranean Plant Protection Organization (EPPO) recommend methods for the assessment of seed health, but these methods are not generally accepted reference methods. Because it is unclear what the most appropriate method is, official organizations responsible for seed health testing and plant quarantine regulations in different countries may use different methods. Obviously, there is a great need for the harmonization and standardization of seed health testing methods between seed testing laboratories of inspection services, plant protection services and industry. However, even generally agreed reference methods can lead to systematic errors, because the criteria and tools used to verify a reliable performance of the test method are absent and variability in test results may occur between laboratories. As a consequence, the same seed lot can be declared healthy by one seed testing laboratory and infected when tested by another laboratory. Therefore, there is a growing need for readily available microbiological reference materials to establish and compare the performance of individual laboratories and to define the limits of inter-laboratory variation.

REFERENCE MATERIALS FOR COMPARATIVE TESTING

Comparative inter-laboratory testing is a very useful instrument for evaluating newly developed test methods in comparison with those which are currently in use, and has definitely led to improved and more standardized methods for seed health testing. However, in the field of seed health analysis there are no internationally agreed reference methods, which can be used as the standard in validation studies of new methods. Furthermore, the use of the same method in different laboratories in many cases gives rise to different results (Van Vuurde and Van den Bovenkamp, 1984; Schaad, 1988). Also, in inter-laboratory studies of food and water microbiology differing results frequently occur (Mooijman *et al.*, 1992). Part of the variability in test results is due to unavoidable random errors inherent in every test procedure. Some differences found between, but also within, laboratories may be due to more obvious factors, such as sample treatment, the heterogeneity of the sample, human factors, technical factors and the processing of information (Franken and Sheppard, 1993). The use of reference samples in comparative tests would provide an additional check of the occurrence of these sources of errors and provide a measure of the performance of both the test and the individual workers. The use of known infected seed samples is not sufficient, because seed samples may be very heterogeneous with respect to the number of infected seeds per sample, the number of pathogen cells per seed, and the quantity and composition of the competing microbiota. Furthermore, samples tend to change in contamination level during storage, due to loss of viability of pathogen cells. Seeds artificially infected with known numbers of pathogen cells also show loss of viability of the pathogen in time, and are not considered to be representative of naturally infected seeds. The necessity of homogeneous reference samples is indicated in the 'Guidelines for collaborative study procedures' of the Association of Official Analytical Chemists (Anon., 1989). Therefore, in order to evaluate the performance and the value of methods there is a great need for (certified) microbiological reference materials that are homogeneous and contain a specified quantity of viable cells of a well defined pathogen strain.

REFERENCE MATERIALS FOR QUALITY ASSURANCE

The implementation of quality assurance schemes in seed test laboratories of seed industry, plant protection services and other governmental organizations will become increasingly important in the near future. In The Netherlands, for example, a quality assurance scheme for seed health testing laboratories has been developed by the General Netherlands Inspection Service for Vegetable and Flower Seeds (NAKG) (see Chapter 30). The goals of quality assurance systems in seed health testing laboratories have been indicated by Franken and

Sheppard (1993): (i) to provide a measure of the performance of the test; (ii) to point out weaknesses of the test; (iii) to provide a permanent record of instrument performance; (iv) to promote a high level of awareness for safety; (v) to assure that personnel are adequately trained; and (vi) to ensure consistency in sample processing and reporting of the data. To fulfil these requirements well described reference methods and reference materials will play an important role.

SPECIFICATIONS OF MICROBIOLOGICAL REFERENCE MATERIALS

For application in microbiological analysis, reference materials can be defined as encapsulated, preserved bacterial preparations containing a specified quantity of viable cells of a well described strain. To be of any use, a reference material must comply with several requirements: it must be homogeneous, stable and representative (Griepink, 1989).

With regard to homogeneity, all samples into which a batch of reference material is divided should have the same number of bacterial cells. For practical reasons the term homogeneity is used to indicate that the variation in the contamination level is within specified limits. The use of a liquid medium inoculated with the target organism will guarantee a homogeneous distribution of target cells, thus allowing the production of reference samples which contain a reproducible amount of target cells. However, such liquids will not be stable, implying that a preservation step is required (Mooijman *et al.*, 1992). Stability is another important requirement. Reference materials should be stable over a defined period of time, i.e. the number of viable bacterial cells present must remain within defined limits during storage. For instance, prepared reference materials should be storable, under defined conditions, for at least 1 year without deviating more than 5% from the average number of viable cells. Only then can reference materials be useful for comparative tests and for regular quality control in laboratories. One should be aware that shipment of the reference materials must also not affect the viability of the preserved bacterial cells.

A pure culture of a representative strain of the target organism should be used for the preparation of the reference materials. Although one could argue that a good reference material should behave as similarly as possible to the normal samples being measured and thus should contain competing microorganisms which are normally present in a seed extract or seed washing, this is hardly feasible. The main difficulties in the development of reference materials will be to achieve the homogeneity and stability desired. The preservation procedure used for the preparation of reference materials is therefore quite important.

PRESERVATION TECHNIQUES FOR PREPARATION OF REFERENCE MATERIALS

Suitable techniques for the preservation and long-term storage of (plant pathogenic) bacteria include lyophilization (freeze-drying) or storage at very low temperatures. These techniques are routinely applied to the maintenance of bacterial cultures where the long-term survival of some cells is more important than the numbers of cells surviving. The quantitative aspect, which is important for the development of reference materials, has not been studied extensively for plant pathogenic bacteria. Preservation techniques which may be considered for preparing reference materials are freeze-drying, spray-drying and storage at low temperature on silica gel or paper discs.

Freeze-drying

Freeze-drying (lyophilization) is commonly used in microbiology. Freeze-dried material can be stored at low temperatures for considerable periods of time. A variant of the lyophilization technique was developed by Gitaitis (1987) using canned evaporated milk as a cryoprotectant. By this method there was excellent retention of viability for more than 1 year. For plant pathogens, the shelf-life of lyophilized Gram-positive coryneform bacteria was shown to be better than for the Gram-negative bacteria (Lelliott and Stead, 1987). A disadvantage may be that the dried material is difficult to mix with other sterile material, which makes it difficult to prepare a quantitatively homogeneous sample.

Spray-drying

Spray-drying is based on drying (bacterial) solutions by means of an air flow with a slightly raised temperature. At the National Institute of Public Health and Environmental Protection (RIVM), Bilthoven, The Netherlands, research has been done with regard to bacterial reference materials for food as well as water microbiology (Anon., 1992a,b). In this research, the use of artificially contaminated spray-dried milk as the basic material for microbiological reference materials was shown to be feasible (Mooijman *et al.*, 1992). The procedure used (see Fig. 31.1) was originally developed for the preparation of a reference material of *Salmonella typhimurium*. A highly 'contaminated' stock, obtained by spray-drying, was stored and when necessary mixed with dry, sterile material, such as milk powder. Reference materials were prepared by filling water-soluble capsules, e.g. gelatin capsules, with small portions of the contaminated powder. The stock of highly contaminated milk powder is prepared once and can be used, if the contamination level maintains within the specified limits, for a long time. Resuscitation of the bacteria takes place by slowly dissolving the gelatin capsules in growth medium at a defined temperature and during a defined time. The capsules containing the *S. typhimurium* reference material could be stored for more than 1.5 years at −20°C without a significant decrease

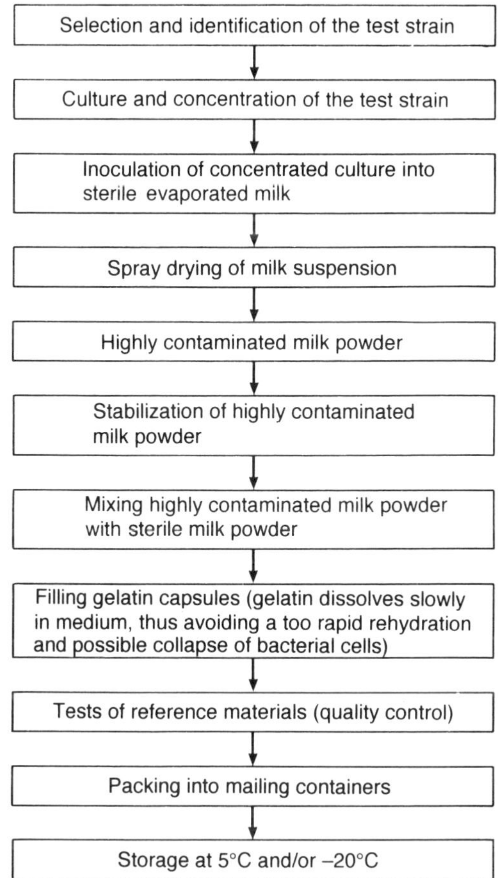

Fig. 31.1. Scheme for the preparation of microbiological reference materials using spray drying. From Mooijman *et al.* (1992).

in the mean contamination level. However, stability of the reference samples could not be guaranteed during transportation. It cannot be predicted whether all bacterial species survive the spray-dry procedure. However, Gram-positive bacteria tend to be better survivors than Gram-negative bacteria (P. In't Veld, personal communication).

Storage at Low Temperature on Silica Gel and Filter Paper Discs

Storage of bacteria on silica gel or filter paper discs is very simple and does not involve a stressful drying procedure as with freeze- or spray-drying. A known number of bacterial cells is added to the silica gel or paper discs, and can subsequently be stored at a low temperature, e.g. –20°C or –70°C. Storage of cells of 13 species of plant pathogenic bacteria in a milk–glycerol solution on silica gel at −20°C even allowed recovery of bacterial cells after 5 years (Leben and Sleesman, 1982). According to Janning and In't Veld (1994), survival of bacterial strains desiccated on anhydrous silica gel and stored at 22°C gives an indication of the stability of the strain in dry, e.g. spray-dried, reference materials. This procedure may therefore be valuable for selecting the bacterial strains to be used in the reference material.

For all these techniques the stability of the product may be improved by the addition of stabilizing components such as glycerol (Gherna, 1994) and the disaccharides trehalose and sucrose (Janning, 1995; Leslie *et al.*, 1995). Storage of the dried material under an atmosphere other than air, e.g. nitrogen or helium, or under vacuum may also be beneficial. Furthermore, besides the preservation procedure itself the thawing and/or recovery of preserved cultures also is a critical step and must be performed under controlled conditions.

QUALITATIVE VERSUS QUANTITATIVE TESTS

In seed health testing a distinction should be made between testing for the presence of quarantine organisms and for the presence of organisms causing the so-called quality diseases. In the case of quarantine organisms a zero tolerance is demanded, meaning that the tests performed simply can be qualitative in nature (presence/absence tests). For this type of test it is important that small numbers of pathogen cells can be detected. The consequence for a reference material for this type of test is that the number of viable cells present per reference sample should also be low. Reference materials for food and water microbiology which are especially designed for such qualitative tests therefore have a low level of contamination of approximately five colony-forming units per capsule.

In the case of organisms causing quality diseases (diseases for which tolerance levels have been set; not subject to quarantine regulations), the test results needed should be quantitative. It is unnecessary to maintain zero tolerances if particular levels of seed infection are unlikely to give rise to disease epidemics in the field under normal conditions. If in the near future generally agreed criteria can be set for tolerance thresholds regarding 'quality' disease organisms, quantitative determination of the level of seed infection will become increasingly important. For this reason reference materials appropriate for quantitative measurements are also needed. Concerning the level of contamination, both low- and high-level reference materials for food and water microbiology are commercially available through the Foundation for the Advancement of Public Health and Environmental Protection (SVM) in The Netherlands.

CERTIFICATION OF REFERENCE MATERIALS

To increase the value of reference materials, a certification procedure for the intended reference material can be followed. According to ISO Guide 30 (Anonymous, 1981), a certified reference material (CRM) can be defined as a reference material for which one or more property values are certified by a technically valid procedure, accompanied by or traceable to a certificate or other documentation which is issued by the certifying body. The standardized procedure that has to be followed for certification is outlined in Fig. 31.2.

PROSPECTS

For use in food and water microbiology, several reference materials have been developed and have been evaluated in European collaborative studies, for example, reference material containing *Listeria monocytogenes* (In't Veld *et al.*, 1995). Gelatin capsules containing spray-dried milk powder contaminated with known numbers of *S. typhimurium* (CRM 507) are now used to validate the detection of salmonellae in food products (In't Veld and Notermans, 1992; In't Veld *et al.*, 1996). Also, reference materials containing *Enterococcus faecium* (CRM 506), *Bacillus cereus* (CRM 528) and *Enterobacter cloacae* (CRM 527) have been certified by the EC Institute for Reference Materials and Measurements, the former European Bureau of Reference.

It is clear that the development of reference materials and certified reference materials will also be valuable in the field of seed health testing. It will promote intercomparisons between laboratories with national and international responsibilities to evaluate and validate the performance of

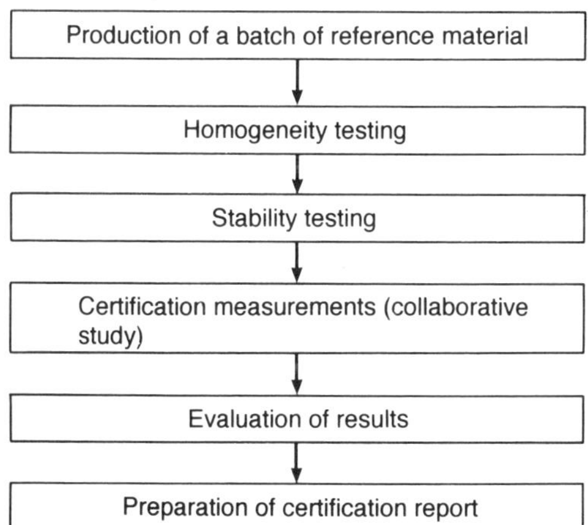

Fig. 31.2. Steps in the procedure for certification. From Anonymous (1994).

currently available seed health tests or of new tests. This will improve the accuracy of test methods and the reliability of test results, and will further international harmonization worldwide by increasing the level of agreement between laboratories. Furthermore, certified reference materials will support the establishment of laboratory quality assurance schemes and provide the means to guarantee that seed health tests are performed correctly to laboratory accreditation bodies.

In the framework of the EC FAIR programme, a project started in 1996, aimed at the development of reference materials to be used in tests for detecting four seedborne bacteria that either may cause considerable financial losses or fall under international plant health regulations, i.e. *Clavibacter michiganensis* subsp. *michiganensis* (causing bacterial canker of tomato), *Erwinia stewartii* (causing Stewart's bacterial wilt of maize), *Pseudomonas syringae* pv. *phaseolicola* (causing halo blight of beans) and *Xanthomonas campestris* pv. *campestris* (causing black rot of cabbage). The development of reference materials for these plant pathogenic bacteria will include research on preservation techniques for the production of reference materials, improvement of preservation by specific treatments of the bacterial cells, testing of different strains, viability testing of samples during long-term storage under various conditions, packaging and testing of the reproducibility of reference materials, challenge tests at higher temperatures and testing the reference material in different seed health testing methods. Comparative tests will be performed by 12 European laboratories to determine the suitability of the reference material for certification. Finally, certification measurements will be performed with those reference materials for which the homogeneity and stability can be guaranteed. Released CRMs should then become available, as test kits, to government and commercial seed testing laboratories. In the future, similar approaches could be applied to the development of reference materials for seedborne viruses and fungi.

REFERENCES

Anonymous (1981) *Terms and Definitions Used in Connection with Reference Materials.* ISO guide 30, 1st edition. International Organization for Standardization, Geneva.

Anonymous (1989) Guidelines for collaborative study procedures. *Journal of the Association of Official Analytical Chemists* 72, 694–704.

Anonymous (1992a) Reference materials for microbiological testing of water (project RM 286). In: *The BCR Programme on Applied Metrology and Chemical Analysis, Projects and Results 1988–92.* Commission of the European Communities, Luxembourg, p. 237.

Anonymous (1992b) Reference materials for microbiological testing of food and feed (project RM 301). In: *The BCR Programme on Applied Metrology and Chemical Analysis, Projects and Results 1988–92.* Commission of the European Communities, Luxembourg, p. 140.

Anonymous (1994) *Guidelines for the Production and Certification of BCR Reference Materials.* European Commission, Directorate General XII, Standards, Measurements and Testing Programme, Document BCR/48/93, 54 pp.

Franken, A.A.J.M. and Sheppard, J.W. (1993) The need for accurate guidelines in comparative testing between seed health testing laboratories. In: Sheppard, J.W. (ed.), *Proceedings of the First ISTA Plant Disease Committee Symposium, Ottawa, Canada*, pp. 143–148.

Gherna, R.L. (1994) Culture preservation. In: Gerhardt, P., Murray, R.G.E., Wood, W.A. and Krieg, N.R. (eds), *Methods for General and Molecular Bacteriology.* American Society for Microbiology, Washington DC, pp. 278–292.

Gitaitis, R.D. (1987) Refinement of lyophilization methodology for storage of large numbers of bacterial strains. *Plant Disease* 71, 615–616.

Griepink, B. (1989) Aiming at good accuracy with the Community Bureau of Reference (BCR). *Quimica Analytica* 8, 1–21.

In't Veld, P.H. and Notermans, S.H.W. (1992) The use of reference materials containing *Salmonella typhimurium* to validate detection of *Salmonella. Journal of Food Protection* 55, 855–858.

In't Veld, P.H., Notermans, S.H.W. and Van de Berg, M. (1995) Potential use of microbiological reference materials for the evaluation of detection methods for *Listeria monocytogenes* and the effect of competitors: a collaborative study. *Food Microbiology* 12, 125–134.

In't Veld, P.H., van Strijp-Lockefeer, N.G.W.M., Havelaar, A.H. and Maier, E.A. (1996) The certification of a reference material for the evaluation of the ISO method for the detection of *Salmonella. Journal of Applied Bacteriology* 80, 496–504.

Janning, B. (1995). Studies on the dessication resistance of various bacterial strains. PhD thesis, Friedrich Wilhelms Universität, Bonn, Germany, 132 pp.

Janning, B. and In't Veld, P.H. (1994) Susceptibility of bacterial strains to dessication: a simple method to test their stability in microbiological reference materials. *Analytica Chimica Acta* 286, 469–476.

Leben, C. and Sleesman, J.P. (1982) Preservation of plant-pathogenic bacteria on silica gel. *Plant Disease* 66, 327.

Lelliott, R.A. and Stead, D.E. (1987) *Methods for the Diagnosis of Bacterial Diseases of Plants – Methods in Plant Pathology*, Vol. 2. Blackwell, Oxford, 216 pp.

Leslie S.B., Israeli, E., Lighthart, B., Crowe, J.H. and Crowe, L.M. (1995) Trehalose and sucrose both protect membranses and proteins in intact bacteria during drying. *Applied Environmental Microbiology* 61, 3592–3597.

Mooijman, K.A., In't Veld, P.H., Hoekstra, J.A., Heisterkamp, S.H., Havelaar, A.H., Notermans, S.H.W., Roberts, D., Griepink, B. and Maier, E.A. (1992) *Development of Microbiological Reference Materials.* Report EUR 14375 EN, Commission of the European Union, Community Bureau of Reference, Brussels, 122 pp.

Schaad, N.W. (1988) Reports of the working groups for comparative tests – working group on bacterial pathogens. In: *Report of the 19th International Seminar on Seed Pathology, Wageningen, 1987.* ISTA, Zürich, pp. 13–16.

Van Vuurde, J.W.L. and Van den Bovenkamp, G.W. (1984). Detection of *Pseudomonas syringae* pv. *phaseolicola. Report of the 18th International Seminar on Seed Pathology, Washington, 1984.* ISTA, Zürich, pp. 8–11.

Quality Control of Seed for Viruses: Present Status and Future Prospects

32

Y. Maury[1] and R.K. Khetarpal[2]

[1]*INRA, Unité de Pathologie Végétale, 78026 Versailles, France;*
[2]*NBPGR, Plant Quarantine Division, New Delhi-110012, India*

INTRODUCTION

International seed trade can be a route for large-scale spread of seed-transmitted viruses. Even low rates of seed transmission in conjunction with secondary spread by vectors can result in the introduction of viruses or their virulent strains into new areas which may lead to the development of disease epidemics. With the increasing global seed trade and exchange, such possible introduction of seed-transmitted viruses needs to be carefully controlled.

The aim of this paper is to highlight the present status on quality control of commercial/bulk seed lots for seed-transmitted viruses, and to outline the future prospects especially in relation to the activities of the ISTA Working Group on Viruses (ISTA-WGV). In this paper the following acronyms have been used for viruses: BBMV (broadbean mottle bromovirus), BBSV (broadbean stain comovirus), BCMV (bean common mosaic potyvirus), BlCMV (blackeye cowpea mosaic potyvirus), BSMV (barley stripe mosaic hordeivirus), CMV (cucumber mosaic cucumovirus), CPMMV (cowpea mild mottle carlavirus), LMV (lettuce mosaic potyvirus), PeMoV (peanut mottle potyvirus), PSV (peanut stunt cucumovirus), PSbMV (pea seedborne mosaic potyvirus), PStV (peanut stripe potyvirus), SMV (soybean mosaic potyvirus) and SqMV (squash mosaic comovirus).

 Seed Health Testing
(eds J.D. Hutchins and J.C. Reeves)

BIOLOGICAL BASIS FOR QUALITY CONTROL OF SEED FOR VIRUSES

Most of the known seed-transmitted viruses are carried within the embryo. Virus in other tissues of mature seed does not play a role in seed transmission (Johansen *et al.*, 1994). Exceptionally, certain stable tobamoviruses contaminate seed surfaces and remain viable in mature testas. Contamination of seed coat surfaces by virus particles results in the mechanical inoculation of seedlings through injuries inflicted by implements during the transplanting process. Such contaminants can mainly be eliminated by heat or chemical treatments (Broadbent, 1963). When sowing seed after such treatments, there always remains a risk of casual increase in soil infectivity during the decomposition of seed coats. Therefore, even for atypical cases of seed transmission, virus-free seeds will be available only through selection of virus-free mother plants or through quality control of seed lots.

METHODOLOGY FOR QUALITY CONTROL

In a seed testing station many seed lots generally need to be tested and low rates of infection have to be detected in large samples. Biological assays, i.e. grow-out and infectivity tests, require a long time for standardization and are laborious and time-consuming especially when working with bulk samples. Recently, sensitive serological detection techniques such as enzyme-linked immunosorbent assay (ELISA) have found their application in testing large numbers of seeds for seedborne viruses (Maury and Khetarpal, 1989). The recent development of the polymerase chain reaction (PCR) technique has resulted in further increases in sensitivity. By using reverse transcription PCR, certain viruses have been detected in seeds, namely PSbMV in pea (Kohnen *et al.*, 1992; Chapter 25, this volume), CMV in lupin (Wylie *et al.*, 1993), BCMV in bean (Saiz *et al.*, 1994) and LMV in lettuce (Chapter 24, this volume). However, its adoption for routine testing needs to be demonstrated.

STANDARDIZATION OF DETECTION TECHNIQUE

The present discussion is mainly based on the application of ELISA for detecting viruses in seeds. During the standardization of techniques, the mode of sample extraction and its dilution limit are parameters that need to be worked out carefully for each host–virus system. Therefore, before the technique can be adopted for routine use the criteria discussed in the following sections should be evaluated.

Detection of Virus in Single Infected Embryos

This aspect has been analysed by ELISA for BSMV/barley (Lister *et al.*, 1981), PeMoV/peanut (Bharathan *et al.*, 1984), PSbMV/pea (Hamilton and Nichols, 1978; Maury *et al.*, 1987), SMV/soybean (Bossennec and Maury, 1978; Lister, 1978; Maury *et al.*, 1983), and SqMV/cucurbits (Nolan and Campbell, 1984). With the exception of BSMV/barley, a large variation of virus concentration has been found in different embryos from the same seed lot.

Some embryos have a very high virus content: SMV/soybean (Bossennec and Maury, 1978), PeMoV/peanut (Bharathan *et al.*, 1984); BCMV/bean (Jafarpour *et al.*, 1979) and LMV/lettuce (Jafarpour *et al.*, 1979; Maury-Chovelon and Lot, personal communication). In contrast, other embryos from the same seed lot have a low virus content: SMV (Maury *et al.*, 1983) and PSbMV (Masmoudi *et al.*, 1994b). These results may correspond to embryos where the virus was only found in the axis and not in the cotyledons, a situation encountered in about 20% of the infected embryos of soybean and pea seed (Maury *et al.*, 1983, 1987).

By diluting each single embryo of a population of infected embryos with an increasing number of healthy seeds, it is possible to determine the probability of detecting one infected embryo in such groups of seeds.

Detection of Virus from Embryos without Interference from Virus in Non-embryonic Tissue

Generally, the percent embryo infection is correlated with the seed transmission rate of a virus in a seed lot. Since the presence of virus in the seed coat does not relate to virus transmission from seeds to seedlings, whole-seed serological assays are not suitable for estimating rates of seed transmission (Johansen *et al.*, 1994). While preparing samples (i.e. extracting embryos) for determining the transmission rate on large number of seeds, it is therefore necessary that viral antigen from non-embryonic tissues (i.e. the seed coat) is not simultaneously extracted, in order to prevent false-positive reactions. Moreover, manual processing of large number of seeds is not possible in a routine test. Four situations have been encountered regarding this problem, and are discussed in the section below.

Virus not detectable in seed testas

Virus was not detected by ELISA in testas of mature peanut seed transmitting PStV (Demski and Warwick, 1986; Xu *et al.*, 1991) and PeMoV (Adams and Kuhn, 1977). In the case of LMV/lettuce seed, good correlation was established between the number of ELISA-positive seeds and the number of infected progeny seedlings. Owing to the difficulty of dissecting small lettuce seeds, it was assumed that the concentration of virus in non-embryonic tissues, if any, was low (Falk and Purcifull, 1983; H. Lot, personal communication). However, these situations could also be interpreted as described below.

Virus present in testas but not extracted

The grinding of whole seeds generally leads to the extraction of viral antigen from the testas also. However, it has been demonstrated in the case of the SMV/soybean system that short blendings of whole seeds using different kinds of homogenizers do not extract the antigen from testas after soaking the seeds overnight (Maury *et al.*, 1985), thus avoiding the laborious step of decortication prior to testing.

Virus extracted from testas but not detected

In contrast with the above findings, it was shown that even short blendings of pea seeds lead to the extraction of PSbMV capsid from testas (Maury *et al.*, 1987). However, it was observed that in testas of such pea seeds, the capsid was partially cleaved during the maturation process. Such partial cleavage of the N- and C-terminal regions of the capsid is well known for potyviruses (Hiebert *et al.*, 1984; Allison *et al.*, 1985). As the N-terminal region is the most immunogenic, it was easy to obtain an antiserum that enabled the detection of PSbMV specifically in the embryos but not in the testas, thus enabling use of whole seed for testing without prior decortication (Masmoudi *et al.*, 1994a).

Virus present in the testas, extracted and detected

In such cases testas have to be removed in order to avoid false-positive reactions. This step can even be complicated by the removal of endosperm (as in the case of barley seeds). It is then necessary to develop a procedure of obtaining clean embryos to precisely determine the seed transmission rate of a seed lot (Hamilton, 1965). Germination of the seed may facilitate obtaining appropriate seedling tissue for assessing the transmission rate.

GROUP TESTING OF SEEDS FOR QUALITY CONTROL

If the percent seed transmission of a seed lot is high, a few seeds can be individually tested by ELISA. However, if the percent transmission is assumed to be low, a large number of seeds need to be tested and an assay on individual seeds becomes practically impossible. Moreover, due to the variation in virus titre in different embryos (Maury *et al.*, 1983) it is not possible to correlate an absorbance value for a group of embryos with the number of infected embryos in the sample. A practical solution consists of dividing a large number of seeds of a bulk seed lot into a number of smaller groups of equal size. The different groups are tested in ELISA as individual composite samples. The decision on the acceptance or rejection of a seed lot can be taken either on the basis of assessment of seed transmission rate or by positioning the seed lot in relation to a level of tolerance. These two alternative approaches are discussed below.

Quality Control by Assessing the Seed Transmission Rate

The group testing method to be adapted for determining seed transmission rates consists of dividing a representative sample of the seed lot to be analysed into N groups of n seeds. The most probable percentage of transmission, $P = 1 - (Y/N)^{1/n}$, can be estimated as a function of the number of ELISA-negative groups (Y). This is demonstrated graphically for a range of seed transmission rates (Fig. 32.1). The precision of the method depends on the magnitude of the confidence interval, for a given level of probability. Curves help to determine the minimum number of groups to be examined; moreover, there is always a gain in precision when the number of groups (N) increases. Other curves give, as a function of the percentage of transmission, the group size limit (limit for n). In such cases N and n would then be chosen as a compromise between economy and the precision required. For routine testing, therefore, a workable group size limit should be determined for each host–virus system (Maury *et al.*, 1985, 1987). It is important to verify that the percentage of ELISA-positive embryos is the same as that of the infected seedlings raised from the same seed lot. Such a correlation has been found satisfactory for SMV/soybean seed (Table 32.1) (Maury *et al.*, 1984). A good correlation has

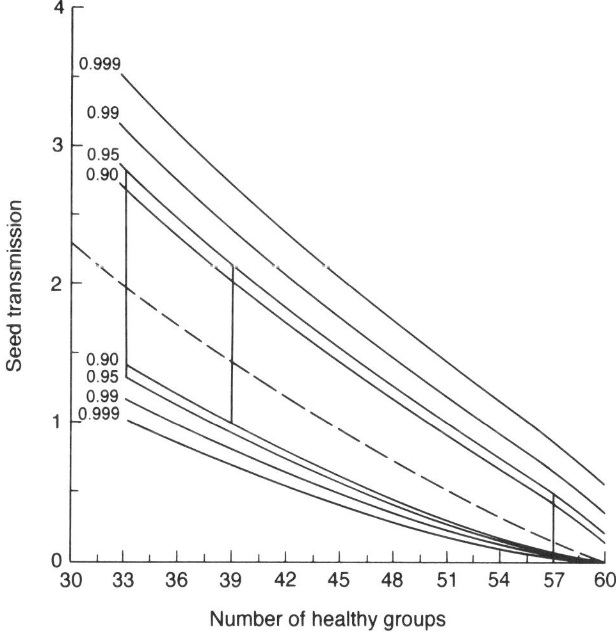

Fig. 32.1. Determination of seed transmission rate as a function of the number of groups of seeds tested. The most probable percentage of seed transmission as a function of the number of healthy groups of seed tested, using 60 groups of 30 seeds (dotted line) and the confidence intervals at the levels 90%, 95%, 99% and 99.9% (Maury *et al.*, 1985).

Table 32.1. Comparative tests on three seed lots of soybean for soybean mosaic potyvirus.[a]

	Seed transmission (%) of different seed lots[b]		
	A	B	C
ELISA			
Braunschweig (Ab1)	0.1 (0.02–0.4)	1.5 (1.0–2.3)	0.5 (0.2–0.9)
Braunschweig (Ab2)	0 (0–0.2)	1.7 (1.1–2.5)	0.8 (0.4–1.4)
Versailles (Ab2)	0 (0–0.2)	2.6 (1.8–3.6)	0.5 (0.3–1.0)
Grow-out	0.1 (0–0.3)	2.5 (1.8–3.2)	0.6 (0.2–0.9)

[a]The seed transmission rate of three seed lots, A, B and C, was assessed using ELISA (60 groups of 30 seeds) with a fast reacting polyclonal antiserum (Ab1) and a slow reacting one (Ab2) at two different locations. This was compared with a grow-out test involving 2000 seedlings.
[b]The figures in parentheses correspond to the confidence interval at 95% confidence (Maury *et al.*, 1984).

also been reported for PeMoV/peanut seed (Bharathan *et al.*, 1984), PSbMV/pea seed (Maury *et al.*, 1987) and BCMV/bean seed (Klein *et al.*, 1992).

Quality Control by Positioning the Seed Lot in Relation to a Level of Tolerance

This alternative approach does not involve the determination of the percentage of transmission of the seed lots. The procedure (Geng *et al.*, 1983) takes into account a non-tolerable level of infection, decided according to its potential of causing economic losses in crop production. A (lower) tolerable level may also be defined according to general usage in quality control. Recently, Masmoudi *et al.* (1994b) extended this approach to the use of group analysis for detecting PSbMV in pea seeds. The procedure consists in testing k groups of N seeds; the decision rule is: 'the seed lot is rejected if at least one group of N seeds is found infected'. The number (k) of groups to be tested can be determined mathematically. This approach enables the use of large group sizes. The number of tests is reduced accordingly. For example, in the PSbMV/pea seed system where inoculum thresholds of 0.1–0.5% can be considered, the number of groups to be tested may be low enough (31 and seven groups of 200 seeds, respectively) to enable a systematic quality control of pea seed to be undertaken (Masmoudi *et al.*, 1994b). This strategy would be chosen for certification of seed lots.

PRESENT STATUS OF SEED CERTIFICATION FOR VIRUSES

To date viruses have rarely been taken into account during seed certification. Different reasons can be attributed or presumed for this, the major being a lack of appropriate methodology in the past and/or a lack of data on the economic significance of viral diseases. The notable example of certification for seed-transmitted viruses is that of lettuce for LMV. Stringent seed certification measures have been adopted due to the high economic value of the crop and the ability of the virus to provoke an epidemic through a very low level of primary inoculum in seeds (Grogan, 1983). While analysing the control of lettuce seed for LMV, we were surprised to notice that, at least in Europe, there is no legal framework for ensuring the quality of seed; companies undertake this control on a voluntary basis. Therefore, some failures have been observed (Lot and Maisonneuve, 1988) and it is evident that certification would give a better guarantee of seed health for internal market and importation. Apart from lettuce, another case of known regional certification is that of barley for BSMV in Montana (Carroll, 1983).

FUTURE PROSPECTS

The renewed interest of the international seed trade in standardized methods of virus detection in seed has been made clear by the recent creation of International Seed Health Initiative (ISHI) (see Chapter 12). Simplification of the methodology of seed certification for viruses calls for an extension of the quality control to even those virus–seed systems where the virus disease provokes lower qualitative and quantitative losses. However, the virus–seed systems to be targeted need to be properly defined.

How do you define the pertinent virus–seed target? A first criterion that could be considered is the existence or absence of a programme of varietal resistance in a seed transmissible virus–host system. Development of resistance leads to the most economic method of control but this approach is followed for only a few diseases. The occurrence of BCMV on beans in USA was considered to be of minor importance after the development of resistant cultivars carrying the *I* gene (Spence and Walkey, 1994). However, severe epidemics due to a necrotic strain of BCMV have recently occurred that could immediately justify seed certification. The introduction of resistance genes into improved cultivars does not hinder the production and exchange of susceptible cultivars, the organoleptic and culinary qualities of which are of prime importance in different regions. In South Africa, BCMV has now recently been included in the Certified Seed Scheme of the Dry Bean Producers Organization (G. Pietersen and J. Serfontein, personal communication).

An important criterion for deciding virus–seed priorities would be the level of seed trade for each crop. However, there are difficulties in evaluating the global seed market (Le Buanec, 1996). Nevertheless, a permanent information channel between ISTA and the International Seed Trade Federation (FIS) is necessary for orienting the work of the ISTA-WGV towards a realistic goal.

Table 32.2. Host–virus systems considered as priorities by the ISTA-WGV.

Crop	Area of production[a] (Mha)	Virus
Beans[b]	24	BCMV, CMV
Tomato[b]	–	tobamoviruses
Pepper[b]	–	tobamoviruses
Broad bean[c]	3	BBSV, BBMV
Cowpea[c]	–	BICMV, CPMMV
Peanut[c]	20	PeMoV, PStV, PSV, CMV

[a]From FAO Statistics, 1993; unknown for cowpea, tomato and pepper.
[b]Crops of international importance in seed trade.
[c]Crops of immense regional importance.

Another criterion would be the importance of the crop as a staple food crop when epidemic outbreaks of viral diseases are completely dependent on the primary seedborne inoculum. However, within each state where such a crop is important, the existence of a seed trade organization is necessary for seed health testing procedures to be used. Besides, even if the level of seed trade is low, the supply of virus-free seed is an important factor in the development of the seed trade.

Taking all these criteria into account, and also the work being undertaken on LMV/lettuce by ISHI, the ISTA-WGV is considering the following virus–host systems as priorities (Table 32.2). An important aspect of the work of the ISTA-WGV for standardization is the definition of antiserum properties, in order to give it an ISTA label as the most appropriate antiserum for seed testing. An accurate protocol using this ISTA-labelled antiserum could then be described in the corresponding Working Sheet. This comparative work will be useful only if there is easy access to the antiserum by seed testing stations and seed companies. Creation of an ISTA Antiserum Bank is critical to ensure this constant availability. This is currently being explored. However, financial support from the industry sectors which will benefit most from this system, is required to develop this approach further.

REFERENCES

Adams, D.B. and Kuhn, C.W. (1977) Seed transmission of peanut mottle virus. *Phytopathology* 67, 1126–1129.

Allison, R.F., Dougherty, W.G., Parks, T.D., Willis, L., Johnston, R.E., Kelly, M. and Armstrong, F.B. (1985) Biochemical analysis of the capsid protein gene and capsid protein of tobacco etch virus: N-terminal amino acids are located on the virion's surface. *Virology* 147, 309–316.

Bharathan, N., Reddy, D.V.R., Rajeshwari, R., Murthy, V.K. and Rao, V.R. (1984) Screening peanut germplasm lines by enzyme-linked immunosorbent assay for seed transmission of peanut mottle virus. *Plant Disease* 68, 757–758.

Bossennec, J.M. and Maury, Y. (1978) Use of the ELISA technique for the detection of soybean mosaic virus in soybean seeds. *Annual Review of Phytopathology* 10, 263–268.

Broadbent, L.H. (1963) The epidemiology of tomato mosaic III. Cleaning virus from hands and tools. *Annals of Applied Biology* 52, 225–232.

Carroll, T.W. (1983) Certification schemes against barley stripe mosaic. *Seed Science and Technology* 11, 1033-1042.

Demski, J.W. and Warwick, D. (1986) Direct test of peanut seed for the detection of peanut stripe virus. *Peanut Science* 13, 38–40.

Falk, B.W. and Purcifull, D.E. (1983) Development and application of an ELISA test to index lettuce seeds for lettuce mosaic virus in Florida. *Plant Disease* 67, 413–416.

Geng, S., Campbell, R.N., Carter, M. and Hills, F.J. (1983) Quality control programs for seed borne pathogens. *Plant Disease* 67, 236–242.

Grogan, R.G. (1983) Lettuce mosaic virus control by use of virus indexed seeds. *Seed Science and Technology* 11, 1043–1049.

Hamilton R.I. (1965) An embryo test for detecting seed-borne barley stripe mosaic virus in barley. *Phytopathology* 55, 798–799.

Hamilton, R.I. and Nichols, C. (1978) Serological methods for detection of pea seed borne mosaic virus in leaves and seeds of *Pisum sativum*. *Phytopathology* 68, 539–543.

Hiebert, E., Tremaine, J.H. and Ronald, W.P. (1984) The effect of limited proteolysis on the amino acid composition of five potyviruses and on the serological reaction and peptide map of the tobacco etch capsid protein. *Phytopathology* 74, 411–416.

Jafarpour, B., Sheperd, R.J. and Grogan, R.G. (1979) Serologic detection of bean common mosaic and lettuce mosaic viruses in seed. *Phytopathology* 69, 1125–1129.

Johansen, E., Edwards, M.C. and Hampton, R.O. (1994) Seed transmission of viruses: current perspectives. *Annual Review of Phytopathology* 32, 363–386.

Klein, R.E., Wyatt, S.D., Kaiser, W.J. and Mink, G.I. (1992) Comparative immunoassays of bean common mosaic virus in individual bean (*Phaseolus vulgaris*) seed and bulked bean seed sample. *Plant Disease* 76, 57–59.

Kohnen, P.D., Dougherty, W.G. and Hampton, R.O. (1992) Detection of pea seedborne mosaic potyvirus by sequence specific enzymatic amplification. *Journal of Virological Methods* 37, 253–258.

Le Buanec, B. (1996) Globalization of the seed industry: current situation and evolution. *Seed Science and Technology* 24, 409–417.

Lister, R.M. (1978) Application of enzyme-linked immunosorbent assay for detecting viruses in soybean seed and plants. *Phytopathology* 68, 1393–1400.

Lister, R.M., Caroll, T.W. and Zaske, S.K. (1981) Sensitive serological detection of barley stripe mosaic virus of barley seed. *Plant Disease* 65, 809–814.

Lot, H. and Maisonneuve, B. (1988) La diversification peut parfois être source d'épidémie. *PHM-Revue Horticole* 289, 33–34.

Masmoudi, K., Suhas, M., Khetarpal, R.K. and Maury, Y. (1994a) Specific serological detection of the transmissible virus in pea seed infected by pea seed borne mosaic virus. *Phytopathology* 84, 756–760.

Masmoudi, K., Duby, C., Suhas, M., Guo, J.Q., Guyot, L., Olivier, V., Taylor, J. and Maury, Y. (1994b) Quality control of pea seed for pea seed borne mosaic virus. *Seed Science and Technology* 22, 407–414.

Maury, Y. and Khetarpal, R.K. (1989) Testing seeds for viruses using ELISA. In: Agnihotri, V.P., Singh, N., Chaubey, H.S., Singh, U.S. and Dwivedi, T.S. (eds), *Perspectives in Phytopathology*. Today and Tomorrow Printers and Publishers, New Delhi, pp. 31–49.

Maury, Y., Bossennec, J.M., Boudazin, G. and Duby, C. (1983) The potential of ELISA in testing soybean seed for soybean mosaic virus. *Seed Science and Technology* 11, 491–503.

Maury, Y., Bossennec, J.M. and Vetten, H.J. (1984) *Report of 18th International Seminar on Seed Pathology, 9–16 July 1984, Washington*, pp. 32–38.

Maury, Y., Duby, C., Bossennec, J.M. and Boudazin, G. (1985) Group analysis using ELISA: determination of the level of transmission of soybean mosaic virus in soybean seed. *Agronomie* 5, 405–415.

Maury, Y., Bossennec, J.M., Boudazin, G., Hampton, R.O., Pietersen, G. and Maguire, J.D. (1987) Factors influencing ELISA evaluation of transmission of pea seed borne mosaic virus in infected pea seed: seed group size and seed decortication. *Agronomie* 7, 225–230.

Nolan, P.A. and Campbell, R.N. (1984) Squash mosaic virus detection in individual seeds and seed lots of cucurbits by enzyme-linked immunosorbent assay. *Plant Disease* 68, 971–975.

Saiz, M., Castro, S., De Blas, C. and Romero, J. (1994) Serotype-specific detection of bean common mosaic potyvirus in bean leaf and seed tissue by enzyme amplification. *Journal of Virological Methods* 50, 145–154.

Spence, N.J. and Walkey, D.G.A. (1994) *Bean Common Mosaic Virus and Related Viruses in Africa*. Bulletin 63, Natural Resources Institute, 168 pp.

Wylie, S., Wilson, C.R., Jones, R.A.C. and Jones, M.G.K. (1993) A polymerase chain reaction assay for cucumber mosaic virus in lupin seeds. *Australian Journal of Agricultural Research* 44, 41–51.

Xu, Z., Chen, K., Zhang, Z. and Chen, J. (1991) Seed transmission of peanut stripe virus in peanut. *Plant Disease* 75, 723–726.

Index